#SciComm – Communicating Science In A Media Age

By Roger A Harris

A guide to understanding the media world
and advice for science communicators.

© 2017 Roger A Harris

D1824064

Contents

Introduction - Why isn't science cool?

Introduction - Why isn't science cool?

Science has an image problem. Sheldon Cooper, Mr Spock, and Professor Frink are exactly what you would expect of a scientist, obsessive, socially dysfunctional and have more than enough knowledge to destroy planets.

Then you have Neil deGrasse Tyson and Brian Cox, two celebrity scientists who display none of the above. And yet because they are scientists, there are people in the world that distrust them. Being an expert isn't cool, knowing a subject inside out and still wanting to know more should be a good thing, and yet to many that means they shouldn't be trusted.

'I believe, therefore I know it's true', like some kind of anti-Descartes. We've all met people who think that vaccines are dangerous, global warming isn't real and for some strange reason that the world is flat. Why do people still hold these to be true, and reject evidence that is shown to them? Could it be that cold hard logic doesn't have a stronger weighting than emotion?

If we allowed courts of law to be run by emotion, no one would receive a fair trial and hopefully most people would be up in arms demanding fair and equal treatment for all. Yet experts are distrusted for using logic to prove why something as obvious as climate change is real and happening right now, or that the world is a globe.

Originally what led me to writing this book was researching the media image of scientists and it's impact on the public. I see these complaints come up all the time in the #SciComm, community on Twitter (hashtag Science Communications), and I've learnt a lot of surprising things because of it.

Scientists themselves have told me that science communication is a problem, but they don't really know how to solve it. I'm not expecting this book to change that overnight, but I am hoping that as someone who studies media, and worked in PR and Marketing that I can help.

And based on some of the communications I've seen, help is most definitely needed. The media stereotype that is often complained about is my starting point for this book, but it's not the problem that many think it is.

Maybe there's a bigger problem that maybe isn't so obvious. Could it be that when scientists are talking to the public, they aren't talking the same language? That's the aim of this book, to give the reader an understanding of the media world, without trying to overload you in theories that don't help you get your message across.

When you see the big names of science communications on television, it's easy to think that they are naturals at this, but so are you. They have the same passion for the subject, but they also have a team of people helping them, producers, directors, and script editors, all working towards the same goal, making a product. Hopefully I'll be able to teach you some of the things that the media teams already know, to help you get your own message across to your own audiences.

1. Understanding the media in a global world

The media image of scientists, the role of the media, Globalisation and Westernisation, Communication Models and Culture.

"The fact that 'Astro Boy' appealed to me as a boy in America was proof that the story and character transcend cultural stereotypes."
Nicolas Cage

1.1 Stereotyping Scientists – Just Stop It Right Now

You might be wondering why a science communication book is starting with the media image of scientists, but there's a good reason for this. I want to start here as I touch on many areas that we'll be looking at in more detail and applying it to communications.

It's also important to remember that when it comes to talking to people, the image you present is important. Everyone knows that first impressions count. The media image will be for most of the public the first interaction with scientists.

It's for this reason we need to start here, so assuming you're a scientist reading this, how damaging do you think the stereotype of scientists in the media is and how often do you see someone complain about this image? My guess is quite often. It comes up every now and then on Twitter in the #SciComm community, and frankly it's time you laid this to rest. This was the subject of my dissertation and the results surprised me.

If you haven't seen the studies by Mead in 1957 and Chambers in 1983, then I'll summarise. What they found is that the at a young age, children are already forming a predefined idea of what a scientist looks like and Doc Brown from *Back to the Future* is it.

Quite a few others had done similar versions of these two studies, and found very similar results and to say I was intrigued to see if it still held true was too tempting to resist. But I also wanted to see what the effect on the science community was as well; after all, you're the ones always complaining about it.

To start with I needed to see just what the public thought, and I focused on adults for this. I created a survey aiming to get a good selection across as many socio-economic types as possible. When they were given a selection of these stereotypical images, only 12.8% said they definitely make them think of real scientists, 53.8% said no, 15.4% maybe and 17.9% said a little.

I need to stress that again, only 12.8% definitely associated these pretend scientist presented in the media with real ones. I do wonder if I had left out the cast of *The Big Bang Theory* if this number might have been lower. But here we have to consider if it was just a case of thinking about pink elephants when being asked not to.

Now the same image went out to scientists in the #SciComm community and the results were quite different when asked if they think it is damaging to the public understanding of science.

Here 42.5% saying yes it is damaging, 33.8% saying a little, 11.3% saying maybe and 13.8% saying no. Here we have the public saying the very opposite of what the science community is saying. Even stranger, out of scientists who say it is damaging, 13.75% were directly inspired by this fictional image of scientists, to become scientists.

One was inspired by Doc Brown, another by

Jurassic Park, although my favourite is the astrophysicist who was inspired by *Stargate SG-1*.

Jurassic Park comes up quite often when scientists complain about this stereotype and here I think it's very important to refer back to a world famous endocrinologist.

Scientists are media dumb.

Michael Crichton, the author of Jurassic Park and the aforementioned endocrinologist (published in the New England Journal of Medicine, and in the Proceedings of the Peabody Museum), gave a talk about this problem back in 1999.

What makes the date more surprising is that Christopher Frayling in *Mad, Bad and Dangerous – the scientist and the cinema,* states that even when *Honey, I Shrunk The Kids* came out in 1989, the stereotype was already looking old, and has rarely been seen since.

If you look at Jurassic Park, the scientists are the heroes; the corporation is the bad guy. I'm struggling to think of a film or television show in the last 20 years where the scientist is not only the bad guy, but even meets the stereotype.

Even in a science communication book that repeated the Mead and Chambers study they also stated the negative portrayal of scientists in the media, but also noted something else; that children had a high level of media literacy.

They understood that what they were describing is a stereotype. If I asked you right now to describe an accountant, would you automatically think of someone who rock climbs, goes free-fall parachuting and weightlifts or would you think of someone in a business suit, probably grey, pallid skin from being indoors too much, and a bit boring?

Stereotypes exist for one simple reason; they create a shorthand of understanding for the audience within a culture to establish a baseline quickly. And that's why when a TV camera comes into your lab, you're asked to wear a lab coat, it creates this shorthand for the audience that you're a scientist.

The lab coat is a key identifying feature of being a scientist for the public. If you do a Google image search for Scientist, you'll see a lot of lab coats, and here this is an example of semiotics. If you've never heard that term before, it means the study of signs and symbols and the lab coat is the biggest sign of science there currently is.

There were others in the past and a walk around the National Gallery in London will introduce you to some of them. In 1645, Salvator Rosa painted a portrait called *Philosophy* and the subject is wearing the scholars gown and cap of the day, using semiotics to inform the viewer who and what they portrait is of. Rosa used the image shorthand of the day, just as in *The Nutty Professor*, Jerry Lewis wore a lab coat, glasses and was a nerd. Then of course there's religious art which even today follows the same codes and symbols used back in the eleventh century.

Stereotypes tell the audience everything they need to know about the subject in question very quickly.

One of these of stereotypes is glasses, which are a sign of learning. This is why Charles Martin-Smith wears them in *The Untouchables* and Danny Glover wears them in *The Royal Tenenbaums*.

Of course they were playing accountants, another group of professionals who are probably more justified than other professions in complaining about their media stereotype in the modern era.

Before I switched to the media world, I worked in a tax office, and for a while they created a cartoon character called Hector the Tax Inspector to encourage people to do their tax returns. He wore grey pinstripe trousers, a black blazer and a bowler hat. I thought he was great and the cartoons, created by a well-known cartoonist, were funny.

Most of my colleagues hated it. All they worried about was people thinking that is what they were like. I'd be surprised if any profession doesn't hate or at least dislike their own media stereotype. You're not alone in not liking being stereotyped, but unlike other professions yours is disappearing.

So if the media no longer uses the stereotype and scientists are more often than not the heroes of film and television, why do scientists still think that the stereotype exists and is damaging?

We've seen that the majority of the public don't think of real scientists when they see fictional ones. And we've seen that their fictional counterparts have influenced real scientists into working in science. So if nothing else, the media image is in fact a positive one. So what is happening?

Well, to explain it, I need to step away from media communications and into sociology (oh the humanities!). First we need to talk a bit about Role Theory, which is quite simple really.

Everyone is an actor playing a role and the easiest way to explain this is a police officer can be a parent, but when they are at work, they don't behave as a parent would. The roles we play sit outside of gender, religion or nationality, which explains why the police behave and talk in the same way, no matter what their gender or cultural background is.

You can observe this in pilots when they give their little speech after take off. "Good morning ladies and gentlemen. Today we'll be flying at 30,000 feet with a good wind behind us so we'll be arriving a little earlier than expected. The weather is good so we're not expecting any turbulence."

Now you've probably just read that in their speech patterns, adopting the role of a pilot. We all know this role and it can even be traced back to one person. Chuck Yeager sounded so in control while breaking the sound barrier, nothing could faze him.

Since then pilots have worked to adopt this role of being calm at all times. It's even transferred to them speaking at press conferences after an accident, using calm measured tones.

So we adapt to the role we are currently occupying and I can't imagine a pilot talking like that at home, although somewhere someone must have done a comedy sketch about it.

And that's an important part about roles, we don't act in them all time, they require a set of characteristics and a context. Pilots learnt the role of being a pilot through two processes, partly through the media and hearing Chuck Yeager, but then more importantly through enculturation; the process through which we learn our place in society and the rules of that society.

Enculturation is happening all the time, and not just in the most obvious way as you grow up. When you start working somewhere new, they have a whole new way of doing the things you already know how to do. And you need to adapt to this enculturation on a much smaller and temporary scale.

Then we need to discuss Self Perception Theory, which is that an individual will self select their attitude based on how they think they should feel in a given situation. Here, Daryl Bem who created the theory posits that our beliefs and feelings come from the overt, rather than within ourselves.

There was an experiment carried out by Festinger and Carlsmith in 1959 where one person had to convince another that a long and boring task was interesting. The one trying to be convinced was then told that the one doing the task was paid either one dollar or twenty dollars to convince them. When they were told it was one dollar, they were more likely to believe that turning a handle for an hour was interesting.

—

When they were told the higher amount, they just thought the other was only interested in the money. The findings of the experiment showed that if someone doesn't hold a strong opinion one way or the other they are more likely to be externally influenced, without understanding the context in which the belief occurs.

Fazio, Zanner and Cooper took the theory further and identified that when you experience cognitive dissonance, or the mental stress of holding two different beliefs at once, your self perception changes to adjust itself and you can be manipulated into a particular opinion or belief. Here is where Malim comes in as in groups, individuals respond to stimuli provided through group interactions.

So if an individual either experiences cognitive dissonance or has no strong opinion on something, they can be swayed into believing something or changing their attitude. Fazio et al also noted that if you do something freely, you'll change your attitude to fit in with that behaviour.

This was seen in an experiment carried out by Freeman and Fraser where they asked people to put a small sign in their windows. When they went back they asked for much bigger signs to be put up, and as the subjects had now identified themselves as helpful, they didn't want to change this attitude so put the sign up.

So how does this fit into the media stereotype of scientists? Well, McCullagh claims that based on studies, it has been shown that the media has limited influence because there is no consistency in the messages being delivered.

So what is having an effect on us?

Culture and group interactions have a much greater influence on the individual as Hogg states it's not the media, but our peers and the society we inhabit. Sociologists no longer believe that the media can influence large audiences as once thought.

This is a good thing as I don't know about you, but I don't want to be anything like the people in some of the reality shows doing the rounds right now.

So why do scientists still complain about the media stereotype? Well, here's a scenario based on my research including surveys and interviews with scientists, through matching the theories I've just talked about and the evidence I collected.

Let's imagine a young person who is regularly consuming media. Now we've seen that the lab coated, glasses wearing middle aged white man with the crazy hair and strange behaviour hasn't been around since the 1990s, so this media image is more likely to be the hero of the story.

As we've seen anyway, the media doesn't influence them because of the lack of consistency in the message of the mad scientist, but there is one consistent message associated to scientists, they are the much more likely to be the heroine or hero of the story. Frankenstein died with his monster.

Because of this heroine and hero, or at least an awareness of them, they decide they want to study in a STEM field, and then they learn the role of being a scientist, the logical structured thought and processes needed of the role, through observations of their professors or supervisors.

Through group interactions and the enculturation process (which are the strongest influence on an individuals behaviour and beliefs), they start to learn that the media stereotype is negative and they go through an attitudinal belief change that they retroactively apply to the meaning of the stereotypical image.

This change happens as they now have new information; they can clearly see that the media representation is (mostly), an inaccurate representation.   Without a strong opinion on the subject when going in, your self-perception can be changed.

And as they are now part of the science community, the belief in the negative effects of the stereotypical image continues and is passed on as part of the cultural history of being a scientist.

As Michael Crichton explained in his speech, and supported by both the evidence and many experts, the media image does not inform the public of what a real scientist looks like or behaves, and explains that they are created this way to facilitate storytelling (Crichton, 1999).

An Astrophysicist I interviewed summed it up perfectly, the public understands that fictional film and television is for entertainment. If they want to learn something, they know that documentaries are for learning.

This media literacy is borne out through talking to the public, and even in research carried out by science communicators who called the media image negative.

People today really do understand that there is a huge difference between real life and fiction.   So basically, relax and enjoy the film, don't worry if they don't show every step of the scientific process as that would be a dull film or TV show.

And when it comes to fictional scientists and character flaws, please take away from this what one scientist (who also thought the media image was negative despite them becoming a biologist because of it), said to me, "Having character flaws is human and scientists are human. So I think you always have to have character flaws for your protagonists. It's important for TV series and films."

"What the mass media offers is not popular art, but entertainment which is intended to be consumed like food, forgotten, and replaced by a new dish."
W. H. Auden

1.2 The Role of the Media

Lord Reith said the purpose of the British Broadcasting Corporation (BBC), was to "Inform, educate and entertain", something they still try to do. Most media companies today have one purpose only and that's making money. It's been said that out of every hundred films, only one makes a lot of money, maybe ten break even and the rest lose money. The industry survives on the big hits.

William Goldman famously said, in Hollywood, "Nobody knows anything", meaning that until people are paying to sit down and watch a film, no one knows if it will make money.

Television is the same, something can be a huge success in one country, but when they try to syndicate it globally, it doesn't work and flops. But the reason they make films and television is to make money.

There are no high ideals behind it. Sure the creators may want to make some interesting points about the human condition, but the people who put the money up are doing so as an investment. You must always remember this.

Again, if you're reading this you're unlikely to be going into this world of traditional broadcast media, although hopefully it will help you look at the newspapers and magazines as an option for science communication. For these too, the reason they exist is to make money.

The local paper in my hometown makes most of its money for the print editions through advertising revenue, not through sales which makeup a third of the production costs. So if the role of the media is to make a profit, how does this help you?

Well, if you can produce valuable content that engages with people, they will be interested. But don't expect to make any money. I was writing for that paper while still studying, and as soon as they asked for my help I asked for my expenses to be covered. They said no.

This was because as I'm sure you're aware, they are now competing with the digital world.

The digital world took traditional media by surprise, and it's taken them a while to catch up, but the two don't really work together perfectly for one simple reason.

Digital media is like life, and as Ferris Bueller said, "Life moves pretty fast", except in the media there's no time to look around. The process for making a film can be years, Television a bit quicker, but how can you create a digital concept, when in six months it's no longer relevant?

Here is where traditional media struggles to catch up, with the amount of money involved a lot of people have to sign off on the idea that may be worthless when it comes to launch date.

There's more flexibility built into digital now, and if you want to see who's really good at it, look at Disney, but so far there's yet to be a media strategy version of the theory of everything.

This hybridisation of the two media is still an evolving process as they try to work out what works and what doesn't. My own field in social media didn't exist in 2010, it's a whole new type of communication and like I say Disney are easily the best. I would love to have a look at their strategy plans.

What makes them excel is that they are

exceptionally good at the one thing that well that all types of media have in common; storytelling. Films, documentaries, news, entertainment shows, even reality shows and The X Factor, they all tell a story about those taking part.

And each story is about an individual's journey. Disney use a well-known story paradigm, and it's worth looking up Pixar's version of this as well, but the original version of it comes from Joseph Campbell in *Hero With a Thousand Faces*.

In this he identified how stories have been told for centuries, following the hero's journey. You'll be familiar with this if you've watched Disney's *Aladdin*, George Lucas used it in *Star Wars*, it's in pretty much every Marvel Cinematic Universe film, even *Game of Thrones* has it.

So how does this help you as a science communicator? Well, make your stories about people.

The reason The X Factor is a successful format around the world is because they tell the story of the competitors, the singing is what drives the show forward through competition, but the viewers come back to find out about those taking part.

In fact, the singing in The X Factor is what Alfred Hitchcock called the McGuffin. This is the device/prop/plot point that drives the story of the film. It is literally the Maltese Falcon, the Iron Throne, the Death Star, and of course the singing on the X Factor. You could argue the prize is the McGuffin as it's what's the contestants are chasing after, but it isn't for the makers of the show.

If you've seen *Ocean's 11*, there's actually two McGuffins. One of them is the cash if the vault (plot A), and the other is making Tess, played by Julia Roberts see that her new boyfriend isn't as good as a person as her ex husband (plot B). In this film the purpose of the story is to get the money and get Tess back, but that's not why you watch and hopefully enjoy it.

You stay watching because the people are engaging and through this you create an emotional bond and invest in the characters through their story.

This is what the media does, it tells you a story about a person, as it's how we connect to each other. We expand on Dunbar's Number through learning about others and welcome them into our own little tribe, even without meeting them.

It's the stories of the people in the media that makes us connect to them, care about them and their progress in life. It's their story that make people care about a celebrity that has never created anything other than having a camera following them around and broadcasted on television. Although in these cases it's a manufactured story created by the programme makers, not a real one. If you haven't seen it, watch *Unreal*, to see just how the media creates a story from nothing.

This is role of the media at its most basic sense, it creates a story that connects us to other people, and the digital world connects us in a way that couldn't be imagined 20 years ago.

But again, with the Internet Dunbar's Number comes back into play. We have a tendency to stick to our tribal groups, and as Hafez pointed out, the web hasn't brought us closer together it has helped to drive us further apart. It's highly unlikely if you're reading this that you haven't seen the responses to 'fake news'; not the type a certain president talks about, but the truly dangerous stuff that leads people down the path of dangerous beliefs.

How often do you see webpages or content that has been shared by neo Nazis, the flat earthers or the anti-vaxxers that hasn't been shared to deride them, and is genuinely their real opinion?

The reason you don't see it very often is because

of algorithms used by search engines, social media and everywhere on the web. These algorithms are becoming better and better at predicting what you're looking for and interested in, to the point that even Google aren't really sure how their search AI does what it does. They only know that it does it well.

What this means if you're a climate change denier who believes the earth is flat and chemtrails are sprayed by the government to control you, well, when you search for information discrediting them you'll find the opposite. We all see what we want to see, and this bias helps to reinforce our beliefs. This happens to all of us, and it's why the scientific process is cold and logical to avoid it impacting on our results.

But we are all human, so when we take the cold logic away......

Even using search engines and the auto complete function can turn up some shocking results. Channel 4 News in the UK, showed some of these in relation to the holocaust, and thankfully Google has done something about that.

But see what happens when you put in chemtrails (*proof, geoengineering, truth*), or climate change (*proof, hoax*) and if you type in the 'vac' of vaccines you get *'Vaccines cause autism proof'*. Once you click on one of these, the algorithm will bring you more next time.

The Facebook algorithm works by showing you what you like and engage with, so again, you'll see the same thing. This also works the opposite way.

If you don't believe in those things, you'll see the opposite so the bubble is still working against you and limiting what media you see and engage with. So rather than increasing our knowledge of the outside world, you just carry on stuck in your bubble, surrounded by other bubbles.

The online bubble may not be new information to you, but you do need to understand it so you can burst other people's bubbles.

*"A lie gets halfway round the world before the
truth has a chance to put its pants on."*
Winston Churchill

1.3 Globalisation and Westernisation– a very quick introduction

While in most cases you won't be trying to do #SciComm on a global scale, understanding this part of media theory helps you when looking at the media as a whole. There's also the fact that in the digital realms your audience could be global.

If you're lucky enough to go viral and have the digital equivalent of 15 minutes of fame, the global media world becomes relevant. Although maybe the term "nine days' wonder", that dates back to Elizabethan times is more appropriate as digital lasts longer.

This section really is to give you an idea of an important part of media theory from a different viewpoint to one you may already be thinking of. For most people Globalisation is a bad thing, but really it's helped raise living and educational standards worldwide. Science of course is a global affair and when looked at this way does change the world.

The Global Village is an idea you've probably heard of and in media terms it's a really important one. Marshall McLuhan is most closely associated with it, and the basic idea is that as the 20th century progressed, we all started to be more connected, see the same things in the media and started to become more alike.

In a way, it's connected to Westernisation as the media world is being pushed more by those of us in the West, more specifically the USA and Hollywood, than from other direction. You may be wondering what this has to do with science communications, but bear with me; this is a journey not a race. Globalisation is nothing new, in a way you could say that the history of the last two millennia is the history of globalisation. The Greeks did it, the Romans did it, the big powers of Europe did it, and of course the church did it.

Now, Globalisation in the modern sense is done through the media and the on-going process of Westernisation. As this isn't a history lesson, I'm not going to waste too much time on the past, but it's important to understand that this is a long continuous wave of how humans interact and look to spread their ideas and ideals to the world.

It used to be by the sword, now it's by the media.

To give an example of how this globalisation happens we only have to watch a film. Hollywood is one of the biggest global media producers, exporting to pretty much everywhere and showing the same product. At one point the US Government even had a department to help make this happen.

When we watch TV we're probably watching American sitcoms, or dramas, maybe not exclusively, but we are watching them. Even when we watch the news, CNN had a huge impact in how we see it, and most news programmes globally look the same now. Global news media is sent round the world and presented to us through our own national broadcasters.

And then of course there's the internet. Here we have not just all the knowledge in the world, but the ability to reach and speak to everyone on the world.

So Globalisation means we're all becoming alike, right?

Well, not exactly.

China limits how many Hollywood films can be released, France and India have their own well established and successful film studios. Films are in a way set in stone, and although the language can be changed, the images and messages of the film can't. But just because the message is being sent, doesn't mean it is the message being received. Maybe, just maybe, someone somewhere is rooting for Darth Vader as they think that law and order is more important than personal freedom.

News media can be changed, you only have too look at how Fox has politicised television news in the US, and newspapers in the UK to see how the same story can be changed to give a different meaning to their audiences.

We can look at Globalisation as being the driving force behind Westernisation but clearly if the news can be changed to meet the expectations of the audience, and how the viewer sees the media reflects how they understand it, does this really help you get your message across?

And what about the Internet, and its role as a globalising force for good? We can all access the same information, read the same articles and speak to each other. But do we?

In our own lives without realising it, we are limiting ourselves to Dunbar's Number, a theory that limits our social interactions to about 120 – 150 people. Or to look at it another way, evolution makes us want to stick to our tribe.

Sometimes we admit new members, sometimes we let others leave, but it happens to us all. You can see this in your digital life in that the average number of friends a person has on Facebook is 120, and you'll know from here that we share friends with strangers, and so our tribes can get bigger or smaller, cross pollinating with other tribes like a digital Venn diagram.

So here we see that the Internet is doing the very opposite of what we think it is. Kai Hafez discusses this in *The Myth of Media Globalisation*. The early Internet as western invention was English speaking, but now it's obvious that countries have the web in their own language, and the news media gets repackaged online for each region of the world.

I saw this happen in my old job in real time with a video of a fishing boat sinking. I posted the video onto the social media accounts, it was picked up by the national and local news agencies and then started to spread globally.

The news story was the same, but different countries reported it differently, and I was able to see the news reports change through a series of Chinese Whispers/Broken Telephone, as it travelled from one location to the next, over a period of four hours.

So what is driving this difference in reporting? Quite simply, culture.

In the western world we tend to think that we have all become the same, and if you think that, please go to France and ask them if it's true. French values are different from those in North America. And in North America, the values are very different in Canada, the USA and Mexico.

Through my own experience in communications in the UK, I can tell you that the culture is different throughout the country, and needs to be considered if you are focusing on local communications or national.

So the global village isn't so much a village, but more a sense of connected villages, that speak to each other, but in different ways. Over that lays westernisation through the media, which is received but not always interpreted the way it was intended too, or even adapted entirely to meet local needs.

"Two monologues do not make a dialogue."
Jeff Daly

1.4 Communication and Models – Models help you communicate

There are many ways of looking at communications, but understanding these models will help you in ways you may not have thought of. Modelling can of course be used in many different ways, and I'm sure you use this type of method in your day-to-day work.

Of course models are only part of the story, as any journalist writing about Leonardo DiCaprio can tell you. That's why this section comes before the 'how to' section; you need to have some understanding of these for application later.

There are a lot of different communication models floating about in the media world, and many more than I could or want to go into here. What you need to know is that the world is a big place, and your message is just a small part of the media sphere competing for attention.

Here I'm hoping that this very small and quick introduction to communication models gives you an idea of just how much is out there. These few are just scratching the surface.

The idea here is to give you a quick understanding of some of the different communication models that are available out there.

The one communication model I see the most in #SciComm, and one you should know about as is the hypodermic model, or syringe theory, which is basically what you see in a lot mass media. Literally the media is being injected into the consciousness of the public.

If you think of a film, or a documentary, the

makers create the message and then put it out there hoping it is received and liked. Once it's out there, that's it, job done. The theory couldn't be simpler, but it's not the only one and this model lacks any finesse. It's the media equivalent of standing on a street corner with a mega phone.

The next stage of communications models and where it starts to have a lot more relevance to you is the linear model.

Berlo identified the linear Sender-Message-Channel-Receiver model, and I really suggest you understand this one as it's an expansion on the Hypodermic Model and gives a lot more flavour and this one requires you to think a lot more about the message you are creating and how you want it to be received.

Sender	Message	Channel	Receiver
Communication	Content	Hearing	Communication
Skills	Elements	Seeing	Skills
Attitude	Treatment	Touching	Attitude
Knowledge	Structure	Tasting	Knowledge
Social System	Codes	Feeling	Social System
Culture			Culture

So here you can see that unlike the Hypodermic model where the sender is just injecting the message to the receiver, Berlo is saying you need to consider your skills, knowledge, social system and culture, break the message down through it's structure and codes and consider the channel you are using to transmit the message.

These channels don't just mean the five literal senses; here you have to consider for example, the unsaid words that convey a message we all understand or the non-verbal messages we see such as body language or the symbols we understand without explanation. Communication is more than the words spoken, here this reminds you to consider the less obvious parts of the message as well as the words spoken. It's about how you and the models work together to create a message through working together.

There's a famous saying about music that it's not the notes you play, it's the notes you don't play. That's the unsaid part of the message that people understand (*eye roll*).

Finally you have considered how the audience receives the message. Do they have the skills, knowledge or even the same culture as you? If not, you have to change your message. There's no Google translate available for this and sadly it requires you doing some research. I'll be talking about culture later on.

This is where the semiotics behind the message begins to come into play. What do the symbols and codes mean to you and what do they mean to your audience.

After all, evolution is just a theory, right?

The symbols and codes of one culture don't always translate across to others and a theory is the perfect example of this. For that symbol, the horse has bolted and you're not going to change it, all you can do is try to explain it in a different way. But Berlo's model will help you to explain it in a different way, through considering the recipient of the message before you create it.

Berlo's model differs from the hypodermic model as here, you're actually thinking about the audience who receives the message, and how they will interpret it. While film is transmitted using the hypodermic model, and not much can be changed for the audience, the marketing of the film doesn't use this model. Here they consider the audience and adapt the message to help sell it. Different territories will often have different trailers, posters and merchandising.

I wrote a case study on the marketing used for *Rogue One: A Star Wars Story* looking at both the US and Chinese markets. In China they don't have the cultural history we do in the west with *Star Wars*, so the marketing was adapted to take this into account,

Then of course for social media, they again use a different model, the next one I'll be talking about as you might notice there's something missing from Berlo's model; feedback from the audience.

Feedback is good and when used effectively helps you become a better communicator. I'll be talking about feedback later on, but for now we'll be looking at it in relation to communication models.

This lack of feedback is a flaw, but not one that stops Berlo's model from being an effective tool for you to use. Right now, we need to look at a model that does include it, so lets have a look at Schramm's model of communication.

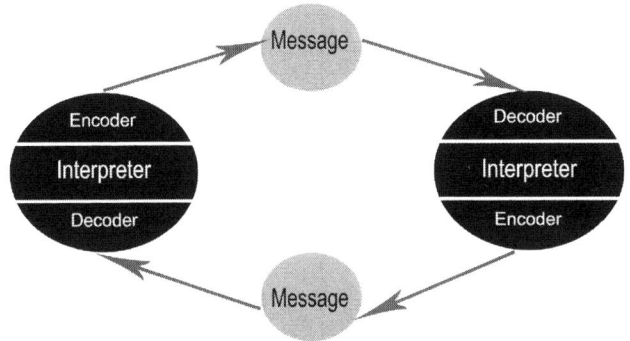

So what does this mean?

Well, the encoder on the left is the sender of the message, and the decoder on the right is the receiver. The receiver takes the message, interprets it and then encodes feedback into a message going back to the original sender. With me so far?

What's important to remember here is that the interpretation process can be seen in the Berlo model where communication skills, attitude, knowledge, social system and culture translates the message into something that can be understood.

You can see this model as a conversation between two people, where you're picking up on not just the verbal and non verbal communication, but all the other aspects at the same time, and adjusting your message as you go along based on the feedback from the other person.

Schramm in his model also highlights something extremely important about communication, and it's something I'm trying to include as much as I can in this book. It's to do with field of experience.

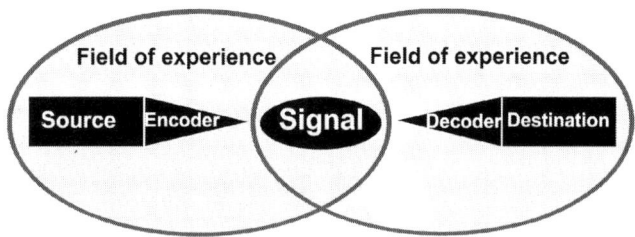

What Schramm was saying is that a message needs to be encoded to meet this field of experience and this can be explained in some simple ways. How would you explain a plane to someone who has never seen one?

They don't understand that heavier than air flight is possible, let alone that people can travel around the world in one. To explain it you would probably have to rely on saying a large bird built by humans that people can fly in, rather than on. This is what field of experience is.

You're encoding the message to fit in with their cultural understanding and experience to make the message understood by the recipient.

Schramm said that successful communication needs this understanding of the recipient's field of experience in relation to your own. Some people may call this dumbing down science, but if they do, their own field of experience is limited. They can't see that not everyone has the benefits of their education.

When we look at both Berlo and Schramm's models together, they look like something we can all work with. But there's another important communication model, the Interactive or Convergence model and this one includes noise.

This model is two versions of linear models such as Berlo's, but one above the other and going in the opposite direction.

Here the feedback isn't continuous, but takes a little more time to come back. You could think of it as posting something on social media, and waiting to see how your audience reacts to it, then adjusting your next message to take their feedback into consideration.

But what about the noise I mentioned? Well, that's the everything else that gets in the way of your message being seen, received and interpreted by the audience. This could be literally anything, the sheer volume of posts on social media, the person sitting next to them chatting away, them thinking about their day etc.

You really can't ignore noise if you want to be a successful communicator in today's world. There's so much of it in our media saturated world, and we all have so little truly free time, that you're in competition with everyone on the planet to get your message across.

And if you're wondering which of these models is the right one for #SciComm, well, it really depends on the platform/channel you're using and what you're trying to say and more importantly, how you're choosing to say it.

What you will find is you'll be using Berlo's model a lot when thinking about how you will create your message, with Schramm's fields of experience for good measure. Picking which models to explain here wasn't easy, I didn't want to swamp you in the details.

It's too easy to not see the woods for the trees when it comes to these, but a basic understanding of them will help you. If you only take away from this book that thinking about your audience before creating any message, I'll be a very happy bunny.

"Cinema reflects culture and there is no harm in adapting technology, but not at the cost of losing your originality."
Jackie Chan

1.5 Culture – We're all the same but different

I touched on this just now with the difference in cultures meaning that some people may not get the message you think you're sending. It's clear that the public don't understand what a scientific theory means, and this is probably the easiest example that shows the difficulty of one culture talking to another.

So in my previous job, this was often an issue where the emergency services would want to say, "A member of the public called us at 21.13 to advise that......", and I'm pretty sure you've seen and heard them say something similar no matter where you live (there's a reason that happens and it's called Role Theory, something I'll come to later).

So as well as the national cultures we all have, there are also other cultures, within national boundaries, and within social groups. This is an area that makes it harder for you to spread your message, as without having any knowledge of these cultures, how can you create a viable message for them?

Dr Ben Goldacre wrote a fantastic book called *Bad Science*, and if you haven't read it I highly recommend it. In it he talks about the media and how they promote a misunderstanding of science to the public. Here we can see one of the problems of one culture trying to talk to another, as he claims that journalists are humanities graduates, not science graduates, who "wear their ignorance as a badge of honour".

I'm a humanities graduate.

So let's say I want to go and talk to street gangs to research them for a study about inter gang communication. The first thing I would do is try to learn a little bit about them before I approach them. If I want them to talk to me, should I wear a suit and carrying my notes in a briefcase, or should I wear jeans?

Well, that answer is obvious, but how should I talk to them to get them to open up? I would mirror their speech patterns but not try and patronise them by copying them completely. I would need them to see I'm not that different from them, but also need them to know I'm not one of them. I would need to maintain a distance, but be close enough for them to allow me into their sub-culture.

To enable the Schramm model of communication to happen and create valuable feedback, I would consider their culture so that worthwhile communication can take place. This is a method used by journalists as well, and why someone who can report on a war, can also interview senior politicians and street gangs. They adapt to the culture of the person they are communicating with.

When I interviewed scientists at the British Science Museum during an evening event, I considered how they would be dressed and matched this (although wearing a NASA tee shirt was a mistake as the public thought I worked for them), and changed my speech patterns to match theirs. Although I was carrying out research interviews, I followed this journalistic style, making it feel like a conversation.

In fact what I was doing was matching their culture to get them to open up, and applying the Schramm model by following the feedback to get to the information I needed.

The process is the same when it's you communicating with them, you consider their culture, and then adapt your message to fit it, using feedback to refine and adapt it to fit their culture norms and understanding.

Culture of course is a very strange thing in that we know what it is, but can't really describe it meaningfully. Robert Pirsig had a similar problem with defining quality in *Zen And The Art Of Motorcycle Maintenance*.

But the most relevant thing about culture is that in communications, it's up to whoever is delivering the message to make it clear to the audience.

Colour is a good example of the risks in these cultural differences. In the west, red means danger (and of course in *Star Trek* we think of all those poor red shirts), but in China red means luck, so if you're planning on sending a global message, how will it work in different countries?

Rogue One: A Star Wars Story used a gold font on black, but in China a white font on black because of the negative connotations the two original colours have there.

Star Wars and *Star Trek* could be claimed to be a culture by themselves that cross the normal cultural barriers, one of the reasons I'm using them as examples here. They've seeped into our collective consciousness and the meanings make sense to us all, making a useful tool for communicators to use due to the differences not just within national boundaries, but reaching across them.

To get a good idea of these cultural differences, we have to talk about Geert Hotstede's theory on national cultural dimensions. Here he looked at each country and gave them a set of values from 0-100, allowing us to compare how one country relates to another. This allows us to adjust our messages accordingly for different countries, or at least be aware of them before creating a message. To give you a quick breakdown of them and as a comparison the scores for both the USA and China. They are:

Power Distance

This about how less powerful people accept how power is distributed in their society. The higher the number relates to how accepting people are in the culture to accept a hierarchical order and their place in society. The lower the number, the more likely they are to strive to create a more even distribution of power and if there are inequalities in power, they want justification for this inequality.

USA 40, China 80

Individualism

This one kind of speaks for itself, but as is always the way with media theories, there's a bit more to it. As well measuring how much an individual can be absorbed into groups, it also considers whether the culture in question considers that the individual should look after themselves, or if a the culture takes a more collective approach. Basically, the higher the number, the more likely the culture is likely to refer to itself as "I", the lower the number, "we".

USA 91, China 20

Masculinity

Masculinity versus Femininity here relates to does the cultural dimension lean more to achievement, heroes, power and rewarding success or does it lean more towards nurturing the members of the society through cooperation, caring for the weak and improving everyone's quality of life (the theory originates in the late 1960s).

USA 62, China 66

Uncertainty Avoidance

There's an often-claimed Chinese curse, "May you live in interesting times." Without going into the accuracy of the origins of the curse, this area relates to how a society feels towards uncertainty and ambiguity. The higher the number the more they want to avoid both of these, the lower, the more they embrace them.

USA 46, China 30

Long Term Orientation
This is about how societies look to the past as a way of dealing with the present and future. A low score here relates to how much they maintain the past at the cost of the new and future. A higher score means they take a pragmatic view of the past and not allow it to hinder the present of future.

USA 26, China 87

Indulgence
Again, this one speaks for itself, is this a society that indulges itself, or does it use restraint. Basically think of yourself, are you someone who eats all the chocolate in one go, or do you make it last a week because that's the right thing to do, but as a society.

USA 68, China 24

Helpfully, you can find the data for each country online using the link in the bibliography or just do a search for 'Geert Hofstede cultural dimensions'. This free online tool allows you to compare one country to another allowing you to see the cultural differences, however you need to remember two things.

It is worth having a look at China and the USA in more detail, or France and Germany. In some ways they are alike, but in many others the differences are poles apart.

You do need to remember that firstly this is both an old theory and old data, and secondly that some scholars have some issues with it. You're more likely to be using it as a guide to remind you that in the global village, everyone is different and you need to adapt your message to fit the cultural expectations of the audience.

These cultural differences are a huge problem in media and communications, especially for advertisers and marketing. If culture creates a barrier, how do we reach out to them without losing the meaning of the message?

This is the Global/Glocal dilemma. The global approach requires standardisation as proposed by Levitt et al, that as globalisation continues to proceed, then this standardisation is the most obvious route forward for communication to all.

But Hofstede shows that different societies have different values and needs. It's easy to look at teenagers in France and Japan and see similarities as a result of globalisation and westernisation, but in fact they are quite literally a world apart. It's been shown that in Japan, those teenagers who look like they are rebelling, will in fact very soon settle down and conform as is the normal cultural values of their society.

These differences are why Coca-Cola thinks local/acts local for advertising. The recent Nintendo Switch game has a different advert for most countries and a search on YouTube for them will show these clear cultural differences where the adverts are tailored for the audience.

Nike of course has the 'Just Do It' slogan and in the western world we all know what that means. Nike adapts this slogan to work glocally, rather than globally. In the USA, they use a 'Righteous Can-Do' with muscular sports stars having a snow day sports game in a parking lot, who quickly form a team who with their can-do attitude overcome the opponents.

In the UK this is adapted to into an ironic dream sequence where Wayne Rooney loses a ball during a game and his life unravels, living in a caravan. But then he snaps out of it, runs the length of the pitch with the ball and wins the game. In India cricketers who are stuck in traffic start a game on the roof of their bus, which spreads out to locals joining in.

None of these adverts would work in the other countries, although maybe the Indian one would work in the UK, but crushing traffic jams are not as common as they are in India, the humour would make sense.

But what about somewhere where the 'entitlement and individualism' of the western understanding of 'Just Do It', doesn't make sense? Maybe somewhere like China?

If you're a global brand and you want to maintain the same message globally, if the sentiment makes no sense, you have to adapt it to keep it.

So In China the slogan means to stay out late, to make friends, to meet your heroes: Just Do It. They've changed the western meaning into an eastern one, adapting to the cultural values of a society.

You can even see this in the *Red Nose Day Actually* sequel to *Love Actually*. It was first shown in March 2017 in the UK and then in May 2017 in the US, and there were a few differences. The obvious one is the change to date that Keira Knightley says, and we'll ignore the extra scene with Laura Linney as it hadn't been filmed in time for the UK version.

Rowan Atkinson reprised the role of Rufus, and in the UK version he was working in a British Supermarket, in the US version he worked in a pharmacy, no big change really, it makes more sense for that to happen.

What is culturally different is the young boy he serves, in both versions played by the same actor, but with different accents. In the UK version he says, "I'll take one" and "alright". In the US, he says, "I'll take it" and "ok". Small differences, but they fit local speech patterns, and clearly culture has been considered for each society.

These differences can be seen in the advertising for *Rogue One* in China when compared to the rest of the world. As they lack the cultural history due to the original films never being shown, what is known everywhere else has to be explained.

For you as a science communicator, the science is the global product that like a film can't be adapted. But as a communicator you need to understand these cultural differences to change how to tell it to individuals or groups through an understanding of the cultural differences.

And it's your job as the communicator to do this; it's not up to the person receiving the message to interpret the message to relate it to their culture. While this so far has focused on national cultures, there are also local cultures, and social cultures.

Within the UK, there are clear cultural differences between geographical areas, and the same is true for the US. My Chinese lecturer would often tell us about different parts of China and what the stereotypical traits were. Every region of the world has these differences within their own borders, so you need to consider these as well when creating your message.

And of course, you need to consider how to translate your own culture as a scientist into something that can be understood by the public. This is Schramm's fields of experience again coming into play.

What you need to search for is the common ground between the two cultures to make both sides come away with a fulfilling experience of understanding.

"Don't text me in that tone of voice!"
Anonymous

2. Communications and Culture

If you want to talk to someone you need to be able to do more than speak their language, you need to understand your audience

2.1 Audience – Just who are you trying to speak to?

So I've given you a quick introduction into the weird world of media communications, but now we need to start talking about how this applies to science communications.

When I returned to university to finish my degree, I had a lot of trouble writing essays because I couldn't write in the academic style anymore.

I'd spent too long writing social media content and blogs that were designed for real people to read. When I say real people, what I mean here is the general public. There's a phrase that has stuck in my mind that relates to journalism and it is 'journalist first, human being second'. Even as a communicator, you are separating yourself from the public. When someone is a journalist, they are taking on the role of communicating with the public. They are no longer a person, but a communicator.

There's a clear separation of roles and I had become comfortable in the role of being a communicator, writing content created explicitly for the public to read and understand. And then suddenly I was no longer needed to write in this style. I had to write in one I hadn't used for a long time, the academic one and needed to change how I saw my role.

It was a common complaint from my supervisor that I wasn't structuring essays and research reports correctly, and it required a lot of effort and work on my part to adapt back to that style. I stuck with it and managed to do ok, although I could tell they weren't completely happy, it wasn't perfect, but at least on the right lines. How I did get there though?

Well, I read a few books, a few websites and more importantly I looked to see how others had done it, I read what they'd written, and not just learned again how to write it, but my last two submissions received my highest marks because of it.

I needed to learn how to write in the style of a culture to communicate with the members who inhabit it. And the reason I had trouble with it was because I had been mainly communicating in the style of a different culture and the expectations and requirements society had placed on it.

This was a problem, but not an insurmountable one.

Now I want you to imagine something, think of a story on local TV news right now, just a short story, one of the filler pieces they use near the end to fill up the available airtime, less than a minute long in total. Now think about how much information they've given you in probably less than a minute; there was an introduction to camera in the studio, an establishing shot that is usually accompanied by a voice over, or the journalist on camera, a short interview and some cutaway shots of what's happening somewhere, followed by a conclusion.

You've been given a lot of details in a very short time, both visual and verbal. Your eyes were seeing the establishing shot which through semiotics explained where they were, and gave a good indication of what the story was about, reinforced by the words spoken in the introduction and the voice over.

The words at the start were probably the least important part in getting your attention, as most communication is non-verbal. The interview tells you who the story is about, and what they say gives more detail. The conclusion will normally just be repetition of what's already been said, and where they are.

Now imagine that same story done in an academic style. I would guess that the same amount of information would be delivered in about five minutes but using a lot more words to explain everything. The academic style of writing is linked to the academic style of thinking.

It's logical, thoughtful, cerebral and basically useless as a form of communication for anything other than academic purposes. If the TV story you imagined in an academic style was done for real, who would want watch it other than academics, and even then would you watch it?

When you deliver a paper or a lecture, your audience is culturally primed to accept it in that style. Even those lecturers you know of who have a lot of flair and that students love their lectures, it's still being delivered in the academic style, but my guess is their papers are pretty much the standard. And that's ok, that's how you need to talk to each other.

Science is about facts and the same is true for reporting the news (or at least as it should be done). So here we have two different ways of reporting facts, but done in two very different styles.

Science is the *What, Why, Where, When* then *Who,* whereas the news is the *Who, What, Where, When and Why.*

The culture of science requires you to take your time walking through each step and explaining them in detail. Pretty much every journalist can cover each in the first two sentences, including a quote, and will then go into detail.

Neither style is better than the other, what's important here is understanding the needs of the audience. Here we have the substance and style that Randy Olsen talks about. Academic communication is about the substance, where news reporting and most forms of communication is style supported by substance.

Quite simply, having expert knowledge of a subject makes it harder to communicate it with the public. The substance of the subject that makes it matter to you, well this gets in the way of the style.

And the greater the substance, the less likely it is that you are able to communicate it effectively to a non-expert.

Originally when I first started thinking about science communications, this style of writing and thinking was leading me down a route where my advice would be that when trying to do #SciComm, you need to separate it between scientists and the public. This makes sense in a way, everyone in science speaks and thinks the same logical way, so therefore they are all within the same cultural boundaries.

But then something happened that really surprised me, and I've seen it a few times now. It needs to be broken down into experts and non-experts.

Let me explain. I was at a Cafe Scientifique event where the talker was explaining Darwin in 45 minutes and I had gone there with a biologist, although I was observing the talk for research.

What surprised me as a non-scientist was the question and answer session at the end. A lot of the questions being asked I knew the answer to, none of which made the evening less enjoyable for me. I asked my scientist friend this and she replied that 'unless you study it, you probably won't know it.'

Now, I'm not in any way a scientist, in fact I left school with barely any qualifications, of which the best result I got was in pottery. I did get a Grade 4 in CSE Science, which was the easiest possible exam to take and even then I didn't do very well.

But I am interested in science (there was a fad in middle school of making clubs for friends, and I made the Science Club), and as such I read it about it whenever I can.

While I have less than no hope of understanding the mathematics for a black hole, or how cell-cell communication works, I do know the important part is the receptor, and that if you cross the event horizon of a black hole you'll end up with a lot in common with spaghetti.

So here is where the cultural part of academia comes in. If you're talking to people in your field, you really are speaking the same language, and carry on doing what you're doing, it's mostly working (just don't forget how long Mendel's theory on what is now known as genetics sat around for before rediscovery).

But when you're talking to non-experts, you need to talk and write in a whole new way.

Communications as we've seen in the Berlo model earlier requires you to think about your skills, knowledge, experience and culture, and how these relate to your audience.

I've done the very same here with the communication models, and the two theories I talked about in relation to the scientist stereotype, I considered my targeted audience and what their relationship and knowledge of the media world is.

I also considered how much time you're likely to have and decided to not over do the theory stuff. Here you only need a basic understanding of it so I can demonstrate my point, then move onto areas that will help you in science communications.

You don't need to research culture in depth to be a good communicator, but you do need to understand the differences between cultures. This is what the next section deals with.

"…..getting to know all about you."
Rodgers & Hammerstein

2.2 Getting to know you - Why culture is important

So why do you need to learn about your audience? Well, there are obvious advantages to understanding a little about who you want to speak to. When you're chatting with a long time friend you have a shared history. You'll have little private jokes only the two of you will understand that have you rolling with laughter, but when you try to explain it to someone else, they give the 'you think that's funny' look.

I only have to say one word to a friend, 'station', and we're rolling around in laughter. Even now I'm smiling as I write this thinking of the joke.

But the fact is, they're right; it's not funny to them. They don't have your shared experience and that's what culture is. You're trying to use the hypodermic model of communication, when really what you need is Berlo's.

Consider your communication skills (*can you explain this to them clearly*), your attitude (*you're trying hard not to laugh*), your knowledge (*only the two of you were there*), your social system (*no one knows you like your friend*), and culture (*how you grew up together*).

Then you need to consider the content, elements, treatment, structure and codes of what you're trying to tell them, and of course the channel you're using which in this case is likely to be using verbal and non-verbal communication through body language.

The last part is of course the repeat of the first. What are the receiver's communication skills (*are they sober or illiterate*), their attitude (*are they getting increasingly annoyed at you laughing at this*), their knowledge (*have they ever seen the bar you're talking about*), social system (*do you really want your partner to know this story*), and culture (*are they even likely to find it funny*).

So just using Berlo's model requires you to know quite a bit about the person you're telling a story to. And of course, the conversation you were having with your friend was using the Schramm model of communication in a feedback loop where both parties were encoding and decoding the messages sent through their shared field of experience.

And in this way, two people can talk about a white rabbit they found in the toilet of a dive bar that they cared for while drunk and woke up the next day to find in their apartment, and then explain to their partners (and that made up story at least has the potential to be funnier than the one behind 'station').

So understanding the needs of the audience means you need to consider a lot of things. In that example, in reality it would more likely be the Schramm model, as the encoders would be seeing the look of confusion and trying to explain why it was funny. And of course we all know that if you have to explain it, it's not funny anymore. Here all those sub headings in Berlo are the interpretation stage and it's not going very well.

So to be a successful communicator requires you to understand your audience, and on social media that's easy. Facebook and Twitter, the big two, give you analytics to work out what's what, if you're a page on Facebook of course (which if you're trying to be a science communicator, you should be).

For social media you should be digging into these details often, not just to see how your latest post did, but also to see who is following you. Look at their ages, their gender, where they are and so on. Facebook is fantastic and gives you so much information it will take a while for you to fully understand how each relates to the other. Twitter tells you what you're audience is mostly interested in (it's normally 'Comedy: Film and Television).

You can also use some free software that a few searches online will bring up many suggestions with varying degrees of accuracy. You could even pay for some software that will give you everything and more, if you have the budget for it. The main thing here is that you start looking at these details and when you do, you can really start to call yourself a communicator.

What if you're giving a speech or a talk? Well, you can still do the same, but now hopefully you'll have more information as whoever is running the event will know who is invited or buying tickets. Is it aimed at the general public? If so, how has it been advertised? Is it through general science magazines, engineering magazines, general interest or lifestyle titles, or on websites?

Wherever the advertising has been carried out will tell you who's it's been aimed at, which in turn gives you an idea of who the audience is likely to be. Through this you can build up an idea of the average audience member.

TV and radio spend a lot of time working out what their average viewer or listener looks like and they adjust the shows to fit these. I once saw the average listener profile for a radio station I listened to, and it broke my heart a little bit.

While I don't remember all the details of it, I do know I'm not woman in her mid forties to late forties with two kids, both in school with a part time job. There was a lot more detail than that, and of the eight or so bullet points I think I only met one of them. The point here being that they had researched who the average listener is, and tailored their shows and the music played on them to meet their expectations.

The reason I mention it here is the age of the

listener is the most important fact for a radio show, as this dictates what sort of music they play. Basically it would be mostly stuff the audience listened to in their teens and early twenties.

Those wonderful years when the world is your oyster and you're partying most nights, with no worries or real plans other than the immediate. We all want those days back, nostalgia is a powerful emotion.

So here they have researched their audience and adjusted the songs play to fit the culture that appeals to them, after all, who wouldn't want to be young again? (If you are young and reading this, understand we hate you for not groaning every time you sit down).

Researching your audience, or at least consider them while you are creating your content. This is time that is never wasted and will pay dividends. Never assume that you're trying to communicate with someone just like yourself.

While peer-to-peer communication is by far the most powerful, you already know how to talk to other scientists. Good communicators learn how to break outside their own culture to find audiences and how to speak to them.

"I never met a chocolate I didn't like."
Deanna Troi, Star Trek: Next Generation

2.3 Empathy to understanding

Ok, so you've done a bit of research into your audience, but how can you turn this into a tool for communication? When I was creating content for social media, and looking at who was following us, I would keep a list of ASL: age, sex and location.

I never had access to education levels and to be honest, I wouldn't have cared about that anyway. We used to write for an audience with a reading age of 15, as when our style guide was written, that was the average. All of the text was crafted for this, unless it was targeted at a particular audience such as children.

To get an idea of this reading age barrier, I've just done a quick search and sadly the average reading age in the UK is 9. The Guardian newspaper has an average reading age of 14, and The Sun, one of the worst tabloids, has an average reading age of 8.

According to the National Literacy Trust, around 16% of the UK population have what is called 'functional literacy', basically, they can understand simple the literacy level of an 11 year old. That's 5.1 out of 53 million people in England alone that require you to use language in a different way, not to mention another 1.7 million adults that are lower than this. This is something you need to consider no matter what part of the world you are in.

I can't stress this part enough, if you're going to communicate something you need the audience to understand it. No one will ever complain about something being easy to understand.

I once created a series of blogs to explain to people

what to check on their boats before going to sea. This wasn't aimed at people who have had a boat for years, it was aimed at a new trend; buying a cheap boat on eBay.

Here the audience had very little boating knowledge and the blogs were created with them in mind. There was also a hope that the established pleasure boater would be interested in it, so for them the hope that it would give them something new to think about. As a communicator I took a very simple approach to this when I wrote the blogs. I put someone else's name on the bottom, from an expert after they had checked it over for accuracy.

This was filled with technical information but at no point did I use jargon, I kept to words everyone knows. They weren't written in a 'you must do this style', as that just turns people off. It was done in a 'it's a good idea to do this now as if the hull is rotten and you're at sea…..' style. This was the friend in the pub giving tips to help a mate out.

I considered the audience, adjusted the language and created the content. Using the experts name also helped to create the 'bloke in the pub' feel, the 'ask John, he's got a boat.'

I wanted to make them as a video with the expert walking around their own boat talking them through it, but sadly there wasn't time. It would have worked a lot better. Did the blogs work? Well, quite a few local papers picked it up and ran them, but more importantly so did the boating press. One of the boating magazines, aimed more at people new to the pastime ran a column that contained every tip, plus a few others that mattered less to us as they were cosmetic.

That's the success you can get when you think about the audience.

So imagine you need to explain the Laws of Motion to someone in their 50's, and this person has never looked at anything to do with science since leaving school. How will you do it without using any science jargon in a way that they will understand?

Look around you right now and what you have to hand is all you can use to help explain it. You're not allowed to write anything down, and you can't try and explain any equations either.

Like I said earlier I'm not a scientist, but the reason I'm using this one is because I once did this at work using a pencil on a desk. I expect now you're starting to see how, as all I needed to do was roll it along the desk. "Things stay still until an external force is applied. It will keep moving until gravity, friction or something else stops it. When I pushed it, it pushed back against me. I know that otherwise I wouldn't have felt it when I flicked it with my finger."

We were both sitting on wheeled office chairs, and I then told them to lift their feet of the floor and I did the same.

I then said, "Every action has an equal and opposite reaction." And gave them a push.

I skipped the second law F=ma as that's much harder to explain in the 15 seconds I had between taking calls, but I explained the main principles quickly and without jargon.

Considering their cultural background, and their own admission of not understanding anything to do with science, I adjusted the message to enable understanding.

Did I dumb it down?

No.

There's no such thing as dumbing down, only understanding the culture and background of the audience, and then adjusting the message to be understood. People who make claims of dumbing down are really saying they are so clever it's not their fault others don't understand.

How would you explain a car to a member of an uncontacted tribe? Would you go into detail about the invention of the internal combustion engine so they can understand the context on the modern car, or start with the Iron Age and how iron was then smelted into steel, vulcanising rubber and so on? How would you explain the differential axle or the carburettor?

Or you would explain it in terms they understand such as a boat that can travel on land, on dry rivers?

I know which one I would choose.

This is what culture is to your audience and the importance of understanding how this impacts on their ability to receive and interpret the message you are sending as seen in the communication models earlier.

Creating empathy for your audience will mean taking Berlo's model to consider their culture and education levels, then looking for Schramm's fields of experience to have clear understanding of the audience's needs from you.

In general terms, in most cases you already know the culture of your audience as you live in it, but as a scientist, you have a higher education level. It's likely that as you've been to university, this is an experience many you're communicating to haven't experienced. Even more so if you've studied at a postgrad level.

Is it dumbing down when you prepare science communications for a school visit explaining chemistry, but not for adults who probably have the same level of knowledge of the subject?

Like I said, there's no such thing as dumbing down, only understanding who your audience is and what they need from you when you are communicating science to them.

3. Testing and peer review

Or how to make sure that your message is a good one through the us of focus groups, informal focus groups, reviewing the effectiveness of the message, and then starting again to make it better.

3.1 "Is this ok?"

One of the things I found out while doing the research for my dissertation that came up on Twitter through the #SciComm community and again during the face-to-face interviews really surprised me. And as a communicator it was actually really shocking to me.

People weren't testing or evaluating the success of their messages.

Let me repeat that.

People are not testing or evaluating the success of their messages.

You can create what you think is the best possible message and graphics that go with it, but without testing them how do you really know it's any good? Without evaluating them, or even setting yourself any targets or looking at the metrics how do you know it is a success?

So lets start at the beginning with this. We'll imagine you know who your audience is, taking into account all the areas I've discussed so far and you've created a great bit of science communication. What's the first thing you should do?

You show it to someone and get their opinion.

Think of this as peer review.

'Have I got the science right' is what you're probably asking another scientist, but they're not your audience. What they want from the communication is very different to what non-experts need from you.

Science is about facts and facts are just a small part of communication, as any politician can tell you. What matters here is what I'll be going into in more detail in the next section about how to write or present your message, what you need to be asking should always be, 'Does this make sense?"

This is the first step in testing your message. You already know the facts are right because that's your job. You're now starting to think like a communicator and understand that the message is the important part.

Randy Olson in *Don't be Such a Scientist* talks about this in how facts are created in the head, but emotions are created lower down in the heart (or even lower, some much lower and much more basic). Do your gut instincts tell you that the message is good?

Take your head out of the message and ask yourself if it appeals to you. Then find out if it appeals to your colleagues, but don't ask them about the science. The facts are still important but you're not trying to communicate details, you need to translate them into the concept.

If they say you're dumbing it down too much, you're on the right path. After all, as Einstein said, "If you can't explain it simply, you don't know it well enough".

The next step in an ideal world is a focus group, but that takes a lot of time and money to do properly, but don't write it off just yet. I would love to do a proper focus group for this book, but realistically it's not going to happen, so I'm doing the next best thing; I'm asking some scientists to read it (and this is good time to thank them publicly for their help with this).

What this does is give you a very good feel for how the message will be received by the rest of your audience. It's no guarantee it will be great, but any indication is better than none. Don't worry about small sample sizes so long as those you're asking are in the targeted audience, as this will reduce any potential sampling error.

I can't say enough how important it is to test your message before you publish, especially in the early stages of you doing any form of communication (or if you're not doing it already). I would never give a presentation before practising it, saying it out loud to make sure I have my timings right. I would never send any message without saying it loud (or quietly in a crowded office), as this is how you spot problems.

One of the first things I was taught many years ago for effective writing in journalism, was to use short sentences, certainly no longer than 15 words, and no more than two sentences in a paragraph. Try to avoid 'hard words', which have three or more syllables as this slows the reader down, and if you're speaking them it slows you down. There's nothing to say you have to stick to these rules, as it depends on your audience and the message you are trying to convey.

By saying your message out loud, you spot the mistakes quickly and you realise exactly why these rules are good ones. Testing is good, rewriting is good and the less words you can use to explain something is better.

Andy Rouse the wildlife photographer said that there are no photos that can't still be cropped a bit better or have a little more editing done on them. George Lucas said a film is never finished, only abandoned, which is just a modern version of Leonardo da Vinci's saying. Another filmmaker said you never finish editing, you just run out of time, and this will happen with your rewriting process.

You'll never be completely happy with it, but by testing it before it's out there and reworking it will always mean it is getting better with each updated version.

So your message is out there, and now you're hoping that it worked, people feel that they've learnt something new and understand the universe a little better. Here, based on what I've learnt is the stage that science communication ends, but hopefully not everywhere.

As a communicator, a campaign wouldn't finish until the report was written (sometimes 300 words, sometimes a lot more). For normal day-to-day content, this would be a week later when I would review the stats. For bigger campaigns it might be six months to a year later.

Lets look at the campaign method first, for the #SciComm you spend a lot of time working on.

So your content is out there after it's been thoroughly tested and improved for your audience, and you're sitting back feeling relaxed and pleased that it is finally done.

Don't.

Now, depending on what platform or format you've used you need to be collecting your evidence of success or failure. Think of it like an experiment where you're not really sure that it's worked out as planned, and you need time to review the results.

For this example, lets say it's a public talk, and you gathered feedback during the event and collected more afterwards. They laughed in most of the right places, they were quiet and asked the right sort of questions that implied they paid attention. This is the point you should ask someone who came with you, or the organiser the most important question you could ever ask: "Did you hear any coughing?"

This might sound strange, but Lord Laurence Olivier once said that theatre acting was all about suppressing the audience's coughs. People who are interested and paying attention don't cough. Sure, if they have a cold they'll cough, but not as much as before.

You'll probably ask them how it went as well, but that's human nature as much as anything. We're social animals and we all want to be loved and appreciated. If you get a chance to mix with the audience and talk to them, you can ask them directly, but remember that bit about us being social animals? They may not want to upset you, so you need to observe their non-verbal language to see if it's true, and also those standing nearby, as their body language will be more telling.

If you video the talk, and are ok with watching yourself on a screen, you'll be your most critical reviewer. Sit yourself down with a notepad and a coffee, it will help you for next time.

If you've written a blog or created some content for online then the evaluation process is instant; sometimes depressingly so, depending on the size of your audience.

The best tip I can give here is to walk away for a bit and have a cup of tea. William Goldman says that in Hollywood, no one really knows what will work, and the same is true online.

Your fantastic bit of science communications that you've tested and everyone loved might just sit there with no one engaging with it. It happens sometimes and makes no sense, so chalk it up to experience and use it again another day at some point in the future.

But hopefully that will be rare for you.

The rest of the time set yourself some targets for the content, this many likes, that many shares, so many views etc. They don't have to be high, but they should be achievable and realistic. Having a target makes a difference, it doesn't mean you've failed when you don't achieve them, it means you have to try to find out why it didn't work.

I used to set two type of targets, the ones I had to achieve and the one I wanted to achieve. The first one, or the official target were realistic and achievable. Basically this would be the one that was low.

The second set of targets I wouldn't tell anyone, this is what I either wanted it to achieve, or what I expected it to do. Sometimes I would tell people these numbers because I wanted to prove that the content was strong enough to perform that well. But mostly I kept that one to myself as you never really know if it will work. This is your personal 'I can do better next time' target or if you're lucky the 'feeling a bit smug right now' target.

The important thing with having something to aim for is it stops you from doing the worse possible crime in communications. Saying something because you think you should. You should be saying it because it's the right thing to do, something I'll be talking about more in section 4.

How you evaluate the success of something is hard to explain, but you'll know when you do well, when you do really well and when you're doing average. It might be that you decide you want to just count likes, shares, but if you do, you need to keep an eye out for serial retweeters. These are the people who follow you and will always do this, no matter what you post.

Deciding whether you count these as part of your

normal total or not is something to consider. These likes and shares are a good indicator, but photo views, link clicks, profile clicks and all the other metrics are much better. Likes and shares help more people see the content, but these are where people are staying and getting involved.

One thing that always surprises me is what people are really interested in, and your stats and metrics will help guide you in this. If you're only communicating to scientists rather than the public, you'll find out quickly enough through this instant digital feedback.

There will always be content that will surprise, it's just one of those things. You won't be able to explain it and not find any reason for why it was a success.

One that happened to me was one of those throwaway types of contents posts that were relevant to the business and kept on hold for when there wasn't anything else happening. I didn't think it was anything special.

As an emergency service, we obviously needed vehicles, and a new delivery had arrived but not in the usual colours. So I grabbed a few of the photos, made a quick collage in Photoshop and posted it with the text '*It doesn't matter the colour of the truck, if the blue lights are flashing, you know what to do*', or something like that.

It went crazy from the start, and carried on for about four days, and I still don't know why. Sure, there's a valid message there, but really it's nothing that the public weren't already aware, and the colours were used by the other emergency services anyway. In Hollywood they call that a '*non-recurring occurrence*' according to Goldman, which is another way of saying no one really knows what will work.

But they keep making *Fast and Furious*, and *Transformers* movies because they do work. They work because they follow the formula based on audience research of what works. The stats and feedback don't lie, no matter what the critics say. Trust them, they are your friend.

It's just that sometimes you get a movie like *The Full Monty* who's success can't be explained in the usual manner, but there's a lot less of them. When it does happens, enjoy it, don't try and over-analyse it, you'll go mad trying to work it out. I'm still trying to work out why those pictures of the vehicles worked so well several years later.

What you want to look for are patterns in engagement and then repeat it, but not in the same way. You'll be tempted to but that that's lazy communication, and it bores the audience.

Say you have a video posted on YouTube and on it gets great engagement on Twitter and Facebook. You can't just post it again the next day, and the day after. In fact, you have to sit on it for a while before you use it again or you'll drive your audience away in boredom. What you need to look for is what they liked about it.

Did they like how you wrote the text that made them click the link, or did the preview image appeal more? It was probably the image as the eye is drawn to them more than text, but maybe the words helped reinforce the image. Maybe those who engaged responded to it emotionally, or was it intellectually?

You'll have an idea of these answers based on the content and it will help you next time, although maybe it was the subject matter that appealed to them more. You can only find this out by keeping a record of what has happened in some way.

Like I said, I kept a piece of paper with a list on, and I would look at the stats and compare them to each other to find a pattern. As it was my job, I would spend a lot of time doing this, but it is is unlikely that you'll be able to. This is why you set yourself targets for each piece of content.

Targets will allow you to instantly know if something worked or not, and maybe for the ones that you think will do well you set a higher target for. Then you need to evaluate the success or failure in some way, whether it's looking at the stats, asking people both before and after publishing what they liked. If you're doing a talk you can hand out a simple tick box survey, but you need to know how you're doing.

Without doing this you'll never get better at communications. And just to make you feel a bit better, this step never stops; it's like 'Painting the Forth Bridge'*.

*This is a cultural reference that may not mean anything to you, and as you have to consider these differences in the sending of your message. I'm making you look it up to understanding the problems in one culture speaking to another.

"Let's start at the very beginning,
A very good place to start,
When you read you begin with A-B-C,
When you sing you begin with do-re-mi"
Rodgers & Hammerstein

4. Practical advice for communications

Introduction – this is what you're here for.

So now you're reading this, I know one of two things about you. There's a good chance you've either flipped to these pages to start here, or you've read the book to here. If you're in the second grouping, you can skip the next paragraph.

If you've skipped to these pages, you will need to read the previous parts or this will not make sense. This isn't an academic book where you can pick and choose what you read. A famous endocrinologist said that scientists are media dumb; you need to understand this world if you want to be a good communicator, let alone a good science communicator.

OK, now we're all up to speed I can talk a bit about what I'll be going through here. I'm going to start with writing your message and I'm going to talk about more than just writing it for a blog. This will include writing a feature article (longer than a blog) and writing for video (a whole new skill). The second is becoming more important these days, something I will talk about when we get to the social media section later on.

Then I want to talk about talks, as in presenting science to the public and how to deliver your talk. I've been to a few of these now and I think I can help you (remember, you already know how to present real science to scientists, but these skills will help there as well).

From here we will move into working with video and how to do #SciComm there, and then into my old speciality, social media. My new one of course is #SciComm.

The last section is about saying the right thing at the right time, as there's nothing worse than seeing something that would have been relevant a week ago when it was in the news, and now no one cares anymore.

One thing you may notice as a recurring theme through the next sections is that I never talk about the details. There's a reason for this and the same will go for you with your science communication.

What these sections are about is the concept of communications, not the details. You will never learn how to be a great communicator from a book, but you can learn the concept of being a communicator.

For you as a science communicator to a non-expert audience, this is extremely relevant to you. In the short time you have to get their attention, details do not matter, the concept does.

God is not in the detail, after all, the bible is pretty thin on details.

> *"The road to hell is paved with adverbs."*
> Stephen King

4.1 How to present your message in writing – and be interesting

I hate writing in the academic style, it's dull, constantly goes backwards to the start over and over again, takes forever to get where it needs to be and when it finally does you've forgotten most of the details that backs it up.

But it has its place, and that place is academia. If you want to speak to real people, by which I mean people who aren't scientists, the general public. I know scientists are real people, but here you're acting as a communicator, not a scientist, or a member of the public anymore. It's part of role theory we were looking at earlier.

I don't want you to think as a scientist when you're writing; I want you to think like a communicator. Here the purpose is to take the facts you learnt as a scientist, translate it into a workable concept, and then explain that concept to a non-expert audience. That's what this section is about; taking a concept and writing something for the audience you've researched.

I mentioned earlier about the number of words in a sentence. We'll we're going to expand on that in a different way. How much can you say in a 15 word sentence and how many facts can you get across?

'Usain Bolt broke the 100 metres World Record last night in 09:58 seconds.'

Thirteen words with four facts. *Usain Bolt, World Record, last night, 09:58.*

The word *seconds* is redundant, but you need to explain the context somehow, as although you would hope they understand you can never be sure. In reality the time of the record would be moved into the second sentence as it's a supplementary fact, not the main one.

Now what I want to do is start with a simple writing structure to start with before going into more detail. The one were going to be looking at is the one you see used by newspapers for reporting the news.

'A fire broke out yesterday evening in the High Street.'

Ten words with three facts. This a better sentence structure because it only has three facts in it and they are ordered in a specific way. What I've done is place an importance value to each fact, then placed them in a particular order.

1, Fire
2, High Street
3, Yesterday

For important messages like this the most important is obviously the fire, then the 'where' and finally 'when' (if you were tweeting this as it was happening right now the order if different).

What you'll notice is I've structured it in the order 1,3, 2 ordering.

The reason for this is how you absorb a message, not to mention that for a news report it makes more sense this way, and flows better. You could easily change the order and it would mean the same, but it works better this way. The flow of the writing is just as important as the structure and style, as it leads you from one part of the story to the next. The three facts are the substance of the message and the flow, or how they are presented is the style of the message.

You've also learnt everything you need to know in the first sentence. If you want to know more you can read on, if you're not really interested, you won't.

In news writing they call this the inverted pyramid and understanding this structure will help your writing. So lets go back to the fire in the high street and carry on with this article.

A fire broke out yesterday evening in the High Street. The fire and rescue service evacuated the building and said that 'no one' was hurt.

Station Officer John Smith said, "Preliminary investigations are that it was an electrical fire. Thankfully residents were alerted by their smoke detector and called the fire service."

The fire happened around 7:30pm when one of the residents, night shift worker Sarah Collins was woken by her alarm.

Sarah said, "I was in bed asleep when the alarm woke me up. It took me a few moments to realise it was in the flat above, but I called the fire service and they told me to get out, just in case."

Sarah said that it could only have been 'five minutes' before the fire service arrived. They broke into the flat, and the small fire was quickly put out.

Station Officer Smith said, "This goes to show how important smoke detectors are. They really can save not just your own life but also your property. We are asking people to check the batteries on their alarms."

So there's a short news article of 180 words that shows the inverted pyramid style of news reporting. At the start there's a lot of information, and as you get further into the story, there's less and less about the fire until you end up with a quote giving a safety message that is only just about linked to the story, and gives no new information.

News writing is a simple structure, so now I want

you to write a very quick one, not even as many words as this if you want (you can even just write it in your head).

Write a quick story in this style about your day. Imagine how you would describe it in the inverted pyramid. What's the most important part? How many facts can you get into a sentence using the 1,2,3 structure from earlier?

I don't want you to spend a lot of time doing this as in most cases you're never going to be writing in this style. It is an important style to understand as a way of learning a very simple writing structure that's in regular use that almost everybody recognises.

When I was being trained in this style, I can't tell you how many I wrote each week, but it was a lot. The lesson here is that to be able to write in a new style takes practice, and a lot of it. It's why you need to show what you write to someone in your intended audience for the feedback you will need to become a better writer.

The inverted pyramid is a simple structure, and in this case you'll not likely to be using it to communicate science, but having an understanding of it will help you when speaking to journalists who are reporting on your findings.

Now I want to discuss features writing. These are the articles you read in newspapers and magazines where they go into more depth and more detail about something. It might be a news subject or a general interest piece, but again, they follow a structure. In news writing the beginning, middle and end are all in the first two sentences, everything else is just filling out each stage.

Feature writing has a clear beginning, middle and end and stays as far away from the inverted pyramid as possible. With this style of writing you're introducing new information in a steady stream to create a narrative in storytelling.

And the most important parts are the beginning and the end. The introduction needs to entice the reader into wanting to know more. If you think of the start of the news articles, you're giving all the main information in one go. For a feature you want to tease that out.

A lowly clerk in a patent office in Switzerland changed the world.......

What's the matter with your energy? Well, as Einstein proved they are both the same really.

The apple never falls far from the tree, but one apple fell so hard that it helped NASA get to the moon.

Here I'm giving a bit of information that hints where I'm going, but doesn't really tell you anything. We're stepping a long way from the science you want to communication and into the world of art, at least at the start of the article.

Communication is a science, you've seen some of the theory behind it (I've only scratched the surface), but it's a science applied through the arts. News writing is cold hard facts presented in a structure that places importance on the delivery of facts and gives each fact a weighting. The art of features writing is in how the facts are presented to the reader.

Feature writing is much more of an art. Give any two journalists the same news story and they will turn out an article that looks almost identical.

Ask them to write a feature piece about something and you'll get as many different versions as there are journalists. This is the art side of media communications and it takes time to learn this through repetition and practice.

If you go on a one week, one day or one afternoon course for science communication, you are not any better than before you started. Sure, you'll come away with a few ideas, but you are not an expert. If you think you are you are suffering from the Carnoustie Effect.

Carnoustie is a golf course in Scotland that was used for the 1999 Open Golf Championship. Locals play there everyday without having any problems. But the winner of the Open was six strokes over par, playing basically like a very good amateur.

The course was windy and the professionals went in with over confidence. Jean van de Velde went into the last hole only needing to be two over par to win. Well, it went very badly for him, but to be fair, he wasn't the only one.

Now the term 'Carnoustie Effect' is being used by the military, stock traders and others to describe those who expected an easy victory or win, but fail.

I want to apply it to science communication as well.

This book is not a quick fix. You need to practice these techniques, test them and learn them.

Anyway, I digress.

So you've written your introduction trying to tease the reader into your story, what next? Well, now you can start your narrative. Let's look at the example introductions again.

A lowly clerk in a patent office in Switzerland changed the world.......

What is this feature going to be about? Am I writing a history of Einstein, a profile, or about his theories and how they changed science? It could be any of them, it could even just be about patents, but the point here is I'm creating a tease, inviting the reader in.

The same goes for the second one.

What's the matter with your energy? Well, as Einstein proved they are both the same really.

Here the options are even wider for where this is going. Now we could go into the speed of light, nuclear energy or a myriad of subjects. It could even be an article about health using pseudo-science is try and make whatever it is they are trying to sell sound like it has real value it hasn't earned.

My favourite though is the last one. This one seems to be a bit more specific, it's either about gravity or going to space.

The apple never falls far from the tree, but one apple fell so hard that it helped NASA get to the moon.

The point here is the intro is not about the subject you're going to write about, it's just an introduction to what you're going to write about.

Now you've written this part, you have to write the hard part, the narrative of the story. There's no easy way to explain how to write it. You might start with the ending and then explain how that ending happened. Maybe you'll start halfway through with equivalent of the eureka moment of Archimedes running naked down the street. Then go back to the start and proving the crown wasn't made of gold, then how that is applied today.

Here the point is you don't have to start at the beginning, it just has to be part of the story you're telling. *A long time ago in a galaxy far, far away* started with episodes 4, 5, 6 then episodes 1, 2, 3, followed by episode 7, *Rogue One* and more to come. You can start anywhere, but wherever you do, be sure that the story starts somewhere and has a narrative.

Again there's no way to get good at this without

practice. It's the only way to get to Carnegie Hall.

None of this means that's there isn't a structure to your feature writing, there is. Remember your introduction? Well, when you finish your feature you need to cycle back round to it in some way.

Lets say you've used the apple and NASA intro then talked about the theory of gravity in an entertaining and informative way, talking about Newton and the laws of motion and gravitational forces. You could end with something like....

The force of Newton's apple changed how we see the movement of the planets and is the reason why we know the path that the Voyager probes will take for thousands of years to come. But the apple still has a role to play with space exploration; astronauts on the ISS just can't get enough of them.

I've used humour, but you can imagine the 800 or so words in the middle now, how the intro gave you a way in to the story you wanted to tell. The ending sums up the whole feature and then flips back to the start, connecting the story that you are telling.

Now you need to write a feature to start to get a feel for how to write them. They're a lot of fun to write in fact, but take a lot longer than the inverted pyramid of news writing. A journalist can knock out ten to twenty news stories a day including research and sourcing the story, but a feature a day is a realistic target, probably with a few news articles as well.

If news reporting is an inverted pyramid, then think of a feature as a column, where the information makes a steady flow in a narrative.

Have a think about writing a feature about your work or research, work out how you would make someone want to read it. If you feel brave you can even try and come up with a good title as well. But remember, you need to practice it to get good at it.

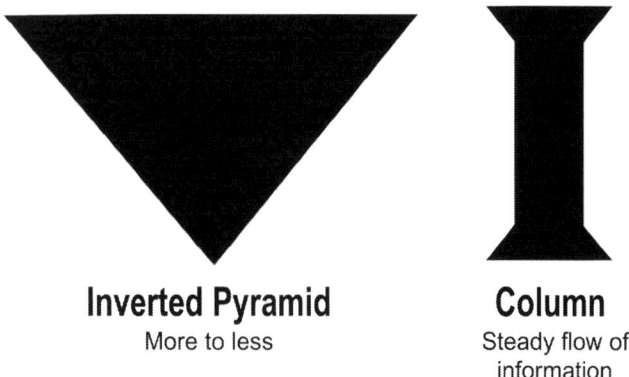

Inverted Pyramid
More to less

Column
Steady flow of
information

Blogs of course are a whole different world of writing. Again, they have a different style and the structure really comes down to your own personal style because blogs are about the person writing it more than the subject being written about.

Let me explain that, a blog is a way for people to learn about the person who is writing it and what they do, rather than the subject they are writing about. Here the reader wants to feel a connection to the writer and their passion about the subject.

I'm really hoping that the whole style of this book is coming across in that style, I want you to understand my passion for trying to help you be better at science communications.

Notice the use of the word 'I' there? I made a connection to you through the use of words to get an idea across. That's what you need to do in blogs.

There are no hard and fast rules for the structure of a blog other than creating a connection to the reader.

Knowing that you're scientists and have a strong connection to logical thought and processes, I will give you a structure to start with. I'm nice like that.

What you need to do is mix the feature and news style into a format that works for you. You also have to accept that most people won't read to the end, just like a news story.

But you've still go to tease them into reading, so you have to mix the feature tease intro with the news style all the facts start. So let's create some right now using the feature intros from earlier.

The apple never falls far from the tree, but one apple fell so hard that it helped NASA get to the moon.

Seeing it's my favourite I'll use this one to turn it into the start of a blog.

Newton didn't invent gravity, he was the first to write down how it works, but how did the apple that fell to earth get man to the moon?

Here I've written both a teaser and what the blog is about. Unlike a feature, you can't be as rambling. People are reading it online, probably on a mobile device so you have to get to the point. When you write your online blog, imagine every reader only has 10% battery power left and that this is your only chance to get them to learn about gravity. You're competing for their attention so you've got to explain it quickly.

Remember we talked about noise in communication, well their battery is about to run out, there's another article of Facebook they want to read and listening to music at the same time is what you're competing with. And it's not a fair fight either.

Blogs are about being quick. 400-500 words is enough, so you have to be clear and concise. Think about the language I used in the news story and how different it was to the feature writing. The blog intro mixed the news and features writing styles, but was a lot less 'flowery' than the later style.

I spent a little time in the British Army, and for giving fire control orders in battle there were two acronyms to remember how to give them. The first was GRIT (Group, Range, Indication, Target), for the order of the instructions. The second acronym was for how they should be said, and it relates to blogs perfectly, CLAP.

Clear
Loud
As an order
With Pauses

So lets go through them. Your blog should be Clear so they can understand it, and kept short so they will keep reading.

Make it Loud and engaging so they will keep reading it. You're competing with the noise of everything in their life and you need to cut through that noise by being engaging.

Make them want to read it by making the title and start as captivating as possible. Deliver it As an order, that is without it being too much like click bait (YOU MUST READ THIS SCIENCE FACT TODAY!). You need to get across the importance of the facts quickly but make them feel you're not delivering a lecture. This is probably the hardest part of your blog, making people want to read it without being too in your face.

And give them a chance to take in the facts, deliver them with Pauses so their brains have time to take in the information in the short time you have their attention. Limit the facts, focus on the concept and think about the 1 3 2, ordering of them.

There are two versions of the same phrase in

writing; the nicer one is called "killing your darlings". The other one is much more brutal and a lot more accurate; "killing your babies".

This is one of the last steps in writing your message and is a part of sub editing it down to the right length. You'll have done this before, but now in a communication setting, you have to be brutal.

Hence killing your babies.

The phrase relates to that bit you've spent a long time working on and you love it, you think it's great, it's funny and explains the science perfectly.

The trouble is, that section no longer works in the larger narrative you've created, this means your baby, well, it's got to go.

Sub editing should be a brutal process and if you can't bring yourself to get rid of the sections you've spent a long time crafting, then you need to think about giving up science communication. That's how important it is.

So lets go through the process for how I approach it for a blog or feature.

Having already done my audience research and considered the style of communication, I write everything I need to write. Once I've finished it, I walk away, I do something else, anything so I'm not thinking about the same piece of communication for a few minutes.

When I come back to it, I read through it, and I read it aloud, this instantly tells me if there are any problems with the flow. As soon as I start tripping up on my words, I know there's an issue that needs fixing.

At this point I will be loosing words and the text will be starting to tighten up, and a flow will be starting to form.

The next stage is to re read it, looking at the way that it's been structured, does it have a narrative and one part lead smoothly into the next.

Once those parts are done, and have been done several times, I stop, make a cup of tea, then start the whole process again. This is all before I would ask anyone to read it.

Here, because this stage would usually be with another communicator, they would do the same as me, except they are helping to make it tighter still, bringing the piece into more focus to help the reader.

If they ask at any point what I was trying to say, then I know I've missed something really important. Here, either they will be able to fix it, or I need to start again.

The tabloid newspaper the sun (it's barely a newspaper and doesn't deserve capitalising), is very good at sub editing. Most newspapers in the UK use news agencies that sell articles of are about 1000 words long. That paper I passionately dislike has an average article length of 300 words.

They do the same for the articles that they do for any written in house. The sub edit the hell out of it. First one journalist edits it down, then the article is passed to another who does the same. When I was told this, the article would normally go through about eight stages until it was passed back to the first in house journalist for a final check.

Then it would go to the Sub Editor who would make more changes and then be ready for printing. In reality these days, there are probably less stages as newspapers no longer have the money. This is how they change a 1000 word article into a 150 word one, and change the reading age from 14, down to that of an eight year old.

So that's a whistle stop introduction in how to write, but you need to remember those communication models from earlier. Which model do these come under?

I'm obviously going to tell you, but here I'm

considering a few things by teasing it out a bit with these meaningless words here.

If you just write what you want using these structures you're using the hypodermic model, but I'm hoping by now you've leant some media theory. A good piece of writing would use the Berlo model. You've considered your communication skills and your attitude, and of course Schramm's fields of experience.

You're learning how to write science communication for the public and your attitude is you want to share these wonderful facts and information with them.

You've got science knowledge by the shedload, and now have an idea how to write. You understand your social system and how that works on the global and glocal sphere. You know your culture and how this relates to how you communicate, and through researching your audience how your message will affect them.

You've broken down the content of your message and its elements, structured it into a way that is understandable using the codes and semiotics for your audience.

Then you've adapted the message to fit the channel you're using, and applying them to the message recipient so they can decode your message through their own communications skills, attitude, knowledge, social system and culture.

But that is only part of the story. Remember the Convergence or interactive model that deals with feedback? That's the one that looks like two of Berlo's models on top of each other but coming back to the original message sender?

Here that takes part mostly before you publish your feature or blog, this is the part where you show what you've written to someone who isn't a scientist. I can't stress how important this stage is.

When I created a series of vlogs for British Science Week, well, the scripts were turned into a blogs to support it. I wrote the scripts and blogs, sent them to the experts and for one of the subjects, they decided they wanted to write their own.

I had no problem with this, but as the communicator, I made it very clear I would rewrite everything they wrote to 'de-clunk it'. I took out the science speak, as it wasn't aimed at scientists and by his own admission he knew he wasn't a writer, he was a physicist.

Through a process of feedback considering all the parts of Berlo's model we were able to craft a message that worked. When you do it this way, you'll have a better chance of getting your science communication right.

Sadly, there are no shortcuts to any of these stages, but they do get easier as you get used to the process. Before long it will become second nature to you.

"Today's public figures can no longer write their own speeches or books, and there is some evidence that they can't read them either."
Gore Vidal

4.2 How to present your message in a public talk

How many times have you given a presentation as part of your work, in a class or as a lecture? A lot I expect, and you probably hated it the first time. That's understandable, you didn't get into STEM to do this.

Well, I hate doing them as well. I hate the pressure, the faces looking back at me, worried I will make a mistake, worried I will get a question I can't answer......The list goes on and on.

The first time I had to do this was when I was 14 as an Army Cadet. The first parade night after getting my first stripe as a new Lance Corporal I was tasked with giving a lecture on the Bren light machine gun. I learnt a lot that night, but one of the most important came in the break.

Our commanding officer had popped his head in to see how I was doing, and asked me to stop by his office during the break. I duly went along and he said he only had one criticism and it was that I was sitting down as I was giving the lecture.

I went back and the rest of the lecture worked much better. Now I know not everyone is able to stand up, and I don't expect you to, but the point here wasn't about standing. It's was about the confidence you get from it. So if you aren't able to stand, think that you are.

I learnt the same thing working in a call centre, if you stand up, you talk differently because you're thinking differently. You're mentally putting yourself in a position of authority and this is what you need to do when delivering a talk.

This is about projecting your personality, which by itself leads to your projecting your voice. If you think about an actor on stage during a play they are controlling the audience not through projecting their voice, but the personality of the character they are playing.

It's through this projection of personality, they lead the audience to where they want to be, to feel the emotions of the character and follow the narrative being created on the stage.

Watch any politician being interviewed, and you'll see how they try to control an interviewer, that's not how you want to be. They are being adversarial, as is the interviewer, and with two sides trying to dominate, neither will back down. That's not where you want to be.

Have a look at stand up comedians and how they behave on stage. They learn how to behave that way through role theory, observing others in the comedy clubs at the early stages of their careers. Ask any comedian about their first gig and they will tell you a story about how they crashed and burned.

Sound familiar?

So the first thing to remember about giving any lecture is you are in charge. This was reinforced to me when again as an Army Cadet, I did the junior leaders course after getting my first stripe. Here we were taught how to give lectures and talks.

Everyone was told before going the usual horror stories that are probably still doing the rounds, and this one wouldn't happen these days, quite rightly. The zeitgeist has changed. We were all warned we would be given five minutes to prepare a ten-minute lecture on the sex life of Ping-Pong balls.

In fact I was given the subject of boots, someone

else had tyres, someone else clouds. The point of this exercise wasn't the subject but the ability to talk with confidence on a subject, even though we had all mentally prepared for the Ping-Pong balls though.

It wasn't completely wasted, as later in school when threatened by a teacher on writing an essay about it as punishment, I just said no problem.

Back to the point at hand. The exercise also taught something else very important for giving talks, and that's spontaneity. Giving a talk is not the same as delivering a lecture, and spontaneity is important here. In a lecture it doesn't matter if you drone on and on and on. They have to listen to you.

Imagine you went to see to see a comedian who just droned on and on and on. Would you be very happy? Some have made it part of their act, but even then, they still display spontaneity, there's a sense that this isn't rehearsed and could go in any direction. It's exciting.

So how do you make is exciting?

Well, you need to get out of your head.

This doesn't mean getting drunk beforehand. If anything avoid alcohol, as I'm sure you're aware all it does is make you over confident. Which of course brings us back to the Carnoustie Effect.

What it means is to stop trying to think with your head and think more with your heart. Randy Olson talks about this in *Don't Be Such a Scientist* (if you haven't read it, it should be the next thing you read).

In it he talks about his acting lessons and the problems he had getting out of his head and how instinctive it was for the others in the class. He gives a great example of how he would knock over a vase during a scene and stare at it before he would say anything.

If you think about how a comedian handles hecklers or unexpected events, then you can see how they work this spontaneity into their acts. I remember seeing Billy Connelly doing this when two people turned up ten minutes late to one of his shows and he started talking to them. He didn't embarrass them, but did give a recap of what had been done so far.

Comedians of course do consider these types of things, especially when they are heckled. Some of the best comedians have a whole part of their act worked around this where they engage with the audience and for this they can't plan. It may be funny, it may not, but the spontaneity of the act is clear to see.

The Irish comedian Dara Ó Briain, himself a great science communicator, does this and then uses what he gets from the audience to further enhance his act, linking them to the funny stories he is already talking about.

So he gets out of his head and into the emotional centres of the body to take the substance and deliver it with a style of his own.

What he is doing is using his head to work out what he wants to say, and then delivering it with the lower centres, mainly the gut, to allow the story to grow. For science communications, you use your head to work out what you want to say, but your heart to deliver it, guided by your gut instincts.

You can of course practise this style of spontaneity. If you look up 'Ronnie Corbett Chair' on YouTube you'll see a scripted joke routine that meanders from what seems like one story to the next. This all helps to deliver context to the original joke being told that without the meandering would be very short.

While you won't necessarily be able to appear as random as Ronnie did from the chair, you can learn from the way he manages to maintain this style while remaining on track.

It may even help you by learning one of these

routines to practice verbatim, to help you deliver your talk in a style that the audience engages with.

Death By PowerPoint

I don't know if you've ever heard this expression, but it's well known in the UK and I expect you have a similar term that speaks for itself. It's a lecture or talk delivered by PowerPoint with no heart or passion where facts are presented on the screen and often just repeated by the presenter.

I don't know how they feel about delivering them but they must be aware that everyone is falling asleep. It's quite possible they are thinking 'this is really interesting', and maybe it is, it's just that I've just never stayed awake long enough to notice.

I used to run a three-hour long training course for social media, using 25 PowerPoint slides with about 40 words of text in it in total, including headings. The longest sentence was a legal warning that had to be included, the rest was just pictures.

I needed to discuss stats and numbers, talk about some of the theories you've seen here. I would discuss with them a lot of the social media section you'll read later. The point being I didn't overload them with something to read while I was talking. A lot of the pictures were cats, because everyone likes a cat picture.

I do the same for academic presentations as well, and the pictures are chosen to back up what I'm saying in a visual way. Most communication is visual, so if you can find a picture that's amusing and supports what you're saying use it. If you can find a video use that, but make sure it's a short video or attention will drift away. Over the last twenty years *The Simpson's* have very helpfully produced a lot of short clips that will probably help you in some way.

There is a danger in using unrelated pictures, but

they do help to get the attention of the audience. They will be waiting for the next one, and only use them on each slide if you have learnt the skill in maintaining the interest of the audience, so use them sparingly to help focus the attention, not detract from the main subject.

The key thing here is to understand that the visual element of your talk will have a greater impact than your words. So you're maybe thinking about how you make graphs and numbers interesting, in which case I've got a lot of bad news for you.

Graphs and numbers are boring. Unless you absolutely have to include them, do everything you can to avoid them. Here's why, when you include them the audience is too busy reading them and trying to decode it than they are in listening to you. I'm going to give you another acronym I picked up in the Army. K.I.S.S. **Keep It Simple, Stupid!**

Does that need explaining? If you are going to include them remember your audience research. Are they able to follow the information you've provided, can they understand simple maths, do numbers scare this type of person? (Numbers scare pretty much everyone, they instantly think there may be a test later).

Remember, what matters here is your audience research, can they handle a lot of information presented in a complex style, or can you make it simpler? Remember, if you can't explain it simply, you don't know it well enough.

If you want to see how to do good a PowerPoint presentation, look up the comedian Dave Gorman. His entire act is a built around given a presentation with his laptop, and he does include numbers and graphs all the time.

But with him, people are going because they know it will be funny. If there's a graph on the screen, they know that there will be a pay off later that will be make them laugh. If you can be funny, you can get away with a lot. If you can be entertaining, you can get away with a bit.

This is the reason I'm saying if you want to be good at delivering talks, watch stand up comedians, they are really good at it.

I went to a talk about black holes in a pub in the UK, Astronomy on Tap, a great idea, people come along for free, have a beer and learn some science.

This particular night there were two talks, one by a PhD candidate, then followed by a well respected astrophysicist. Both talks were interesting, but I was more interested in the styles they were done in. Here we get to talk about style over substance/concepts over details again. Which is more important, the substance that gives a lot of details or the style they are presented in?

If you're giving a lecture, it's substance, as your students and colleagues need that. You're talking to experts, or at least people who will directly benefit from the substance you're delivering.

If you're giving a talk, it's style. Here you're talking to the public or non-experts who want to learn a little, but will gain nothing from graphs and a lot of details.

And here is where both talks failed as neither could decide how they really wanted to give the talk. The PhD candidate had created a PowerPoint that was high on style and low on substance. It was perfectly done to engage. The trouble was they weren't very good at public talking.

They moved around like a comedian would, smiled and were perfectly engaging, but there were constant er's, um's and so on. Quite often they directed the talk at their friends sitting on the side, laughing along with private jokes with them.

This can be distracting for the audience, as who

are they delivering the talk to, their friends or everyone else who was there? It's tempting to want to present to the people you know, and if they are there you may end up doing this. It's everyone else who is important, not your friends and family, they already love you.

Style wise they were close, and I expect that once they are more familiar with delivering a talk in a public setting and learn to overcome the nerves of speaking to the public (with a lot of their peers watching), they will be a great science communicator. I truly hope the experience didn't put them off.

Now, the second speaker again was a mixture of style and substance, but again there were problems. I spoke to them before the talk and asked if they had spoken to their science outreach officer at the university before the talk and they admitted they hadn't. They also said they will in future, and this is a good thing.

Their PowerPoint had a lot of graphs and I spent at least ten minutes looking at it trying to decipher it. I'm extremely interested in science; after all, I do have a grade 4 CSE in it. I've also always been very bad at maths, and was never able to get my head round the twelve times table. So you can imagine how I feel about trying to decipher graphs about black holes.

Their speaking style was great, but they stood still the whole time which meant most of the audience had three choices, look at them standing still, look at the incomprehensible graph on the screen, or look around (which is, all together now: noise).

As with writing, delivering a talk is like writing a blog, you need detail, you need to tease them, and you need interesting pictures or videos. And of course you need the skills of the stand up comedian with the knowledge of an academic.

No one said it would be easy, but then that's why you do it or it wouldn't be fun.

So I've talked about the need to project your personality so that you are commanding the audience to look at you, but what else can you do? Well, I'm sure you all know about using cards rather than a sheet of paper. And I'm hoping you only put bullet points on the cards rather than all the text. Personally what I do is write the bullet points in a large text and either spread the cards out in front of me on the floor, or tape a few pages of A4 with large type on them down.

You'll find your own way to do this, but do not keep looking at them. You should have written your talk so that one part flows into the next, just like you would write a feature piece, weaving a narrative in and out of your subject to lead the audience to where you want them to be.

When you're standing on the stage, the worst thing you can do is stand still, and if there's a lectern, don't get stuck behind it. Again, looking at comedians think about what they are doing as they go through their routine.

They are on a stage, sometimes a big one, sometimes not much bigger than a table. But they don't stand still. After all, a moving target is harder to hit! When you are standing up and talking to a friend, both of you are constantly shifting your weight, moving around on the spot.

Because you're having a conversation, how you're talking to each other is fluid, and your movements match this. When you prepare a talk, because you have a set of words you want to say and memorised the order in some way, your body tends to go stiff as you focus your attention on trying to remember what to say next.

When you're talking to your friend and moving

around on the spot, your body language is open as you know and trust them. Here is a common mistake people make during talks. They get nervous standing in front of strangers and they close their body down.

There's a trick that radio presenters use when sitting in the studio and that's talking to the clock. They imagine the clock is a person they are talking to, and it changes how they talk. They voice opens up and they sound natural.

What you need is a clock to talk to when you're standing there, feeling nervous standing in front of strangers, do you remember everything, what about if you get a difficult question, oh no, I've forgotten what I need to say at the start……..

Imagine that the wall, a clock, a person in the audience, anything is you talking to your friend and just relax. But unlike the radio presenter, you can't keep focusing on the clock while you talk. You need to make a connection to your audience and look at them. They become your pretend friend, so that as you talk you're talking naturally.

And talking naturally is the key here, which links back to spontaneity, when you're trying to remember the script, you forget everything else. As I write this, a couple of weeks ago I was on a Hollywood film set. I can't tell you what film yet as I'm planning for this to be released before the film comes out, so I'm limited in what I can say.

But the star of the film had to perform the same actions about 40 times throughout the course of the day. Every time they did it, they had to make it look like it was the first time they were ever doing it. The only way you can do that is to forget what you _have_ to do, and let what you _are_ doing happen.

You know your subject when you stand up in front of the audience to start speaking. You know the order of the words you want to say. You have your bullet points somewhere you can see them. Now forget them, you should only be using them to keep you on track.

If you drift off track while speaking, it doesn't matter, so long as it is relevant and you come back onto the course you need to be on. Ronnie Corbett was the master at this and loved because of it. This is where your spontaneity comes in, where you make sure that no matter how many times you say it, you're saying it for the first time.

Think about when you ring a call centre and they answer the phone, sounding enthusiastic, and when they answer sounding bored. Having worked in a call centre, I can tell you the person answering is bored, no matter how welcoming they sound to you. If a call centre worker can make it sound like they are saying hello for the first time that day, you can deliver a talk that way.

Of course, speaking is just a small part of communication, this is why I'm saying don't stand still. If you're still, then there's nothing to watch or focus on. Sure you've got your slides, for them to look at. But this is about communicating, not teaching.

So you're moving about and using your hands when you speak, but are you fidgeting? I'm a terrible fidget, always have a pen in my hand that I'm spinning around, tapping on my hand or leg when I talk.

If I'm giving a presentation, I make sure my hands are empty. I also try everything I can to keep my hands out of my pockets, especially hooking one hand in my back pocket. It's important that your hands are in the open and can be seen. This makes you more trustable.

Everyone knows the correct side of the road to drive on is the left. This is so that you can use your sword on a horse rider coming the other way if you need to defend yourself. This is why Napoleon changed the side of the road in Europe when he was in charge (this may be a myth, but it's a good story, so let's go with it).

But the sword is relevant here as having your hands in the open is a trust thing. The same goes for holding a pen or something in your hand; they could be a weapon.

It seems such a strange thing, but our lizard brain isn't very clever, and instincts can override logic. J.K. Rowling tweeted about avoiding a sleeping spider for a couple of hours, only to find out it was a tomato stem on the floor.

The instincts of the audience can make them not like you, or at least feel uncomfortable without any real understanding of why. This is why your body language is so important.

There are many books available on this, but quite simply, all you need to do is stand open and think 'smile'. That means don't cross both your arms, and try to avoid any barriers between you and the audience. Even just thinking of smiling you will start to open up and project your personality.

You'll know if you're not getting this right through the feedback we've seen in the Schramm model. You're encoding information, transmitting the message and they receive it, interpret it, and then send you a message back to let you know if they don't understand, or even worse if they are bored.

You've seen this feedback before, you've probably even sent some of it when you were still a student. When you're giving a lecture at university, you can tell if the students aren't paying attention or bored. It's quite likely that you've seen this and decided that they are the ones who need to know it, so you've carried on.

And it that setting, it's understandable. It's wrong, but it is understandable as they are paying for this lecture, so if they don't pay attention, it's not your fault, right? Well, the audience you are delivering your talk to is paying for it, either with money or the their time.

When I was giving lectures or talks and I saw I was losing the audience, I changed something. I would step off the script, suddenly change the pitch of my voice. Once I even moved and stood behind them so I was looking at the screen.

This last one is a good one as it forces the audience to listen to your voice as they lose all of the non- verbal communication from you. It also unsettles the audience a little as they don't know what you're doing, so don't do it for too long, just to bring them back to where you need to be. And make sure you have something on the screen to talk about that is interesting, as there's nothing else for them to look at.

And of course if they start coughing, you need to be as captivating as Laurence Olivier on the stage to stop them.

Q&A

I've included this very short section as a separate area to deal with an important part of most public talks. A public talk follows the Schramm model of communication, you're encoding the message, and getting direct feedback from the audience if you can see their faces, hear them coughing or talking or walking away to get another drink or any of the things that can distract them from you.

The biggest fear you should have is seeing a lot of peoples faces light up when they look at their phone. But the Q&A section will give you a good idea how well you've done.

Let's say someone asks a question that you know

you've covered really well in the talk. How does everyone else react? If they all lean in closer, that tells you that you didn't explain it as well as you thought. What if they all shift in their seats and look around? Well, you've done a good job, but this one person needs a bit more help.

These are things you will learn from doing your academic presentations, but this time you have to make a choice. How much time do you spend going over it again without losing everyone else in the audience?

Only you will know that as you're the one there, so of course you need to try, but you might need to offer them another way of finding out so you can move on.

I was at a talk about evolution and Darwin, and during the Q&A someone asked the question that was effectively about intelligent design.

The questioner claimed not to be trying to cause trouble, they just wanted to understand it. The speaker had taught biology in the bible belt in the US, so this was nothing new to them and how they handled it was beyond exemplary.

What they did was change how they were talking, they adapted the message for the recipient, making it as easy as possible to understand. After the first answer, they then asked another question on the same line, and the audience started to get both amused and annoyed in equal measure.

The speakers post grad students were in fact the worse, laughing quite obviously at them (tell your students never to do that, all it does is make them look arrogant).

The point here is that the speaker was reading not just the audience member who asked the question and changed how they spoke to them, they also tried very hard to get out of the line of questioning to return to everyone else by offering to speak to them after.

The questioners probably were there to try and cause trouble, as they asked too many questions for that to be true and left soon after, skipping the offer of talking more with the speaker later.

The key thing here is you need to learn how to read an audience, and the Carnegie hall part comes in again here. Practice with someone or you'll never get good as a communicator.

That one day or one week science communication course won't help you. If you want to challenge yourself, try that exercise I had as a kid, five minutes to prepare a ten-minute lecture to your friends and colleagues.

Be sure to give them a really difficult subject, they'll do the same to you.

"If it can be written or thought it can be filmed."
Stanley Kubrick

4.3 How to present your message on video

As I write this, tomorrow in the UK is Election Day, and without going into all the stupid reasons it was called or how ridiculous is it, one theme of many (apart from the party in power who called the snap election), is to get young people to vote.

The UK TV channel E4 has been very good at promoting not just registering to vote in time, but also in encouraging them to vote. I'm sure you've worked out by now that E4 is a channel aimed at young people, but enjoyed by many age groups (I like it).

They created a video campaign that is one of the best I've seen in a long time. In it they have actors, TV presenters and comedians slightly older than first time voters, but still young enough to be cool. They all talk about their first time, while looking into the camera. They say how they didn't know what it would be like, how they felt awkward and worried they would do it wrong, how their first time was in a booth with their parents next door and so on. One of them says it only took two strokes and it was over, they'd done it finally and how doing it made them feel grown up.

So lets use discourse to break down what they've done. Discourse in the media sense is how you describe ideas and how they relate to each other in a social and cultural context. For example, if I said 'cat', you're likely to think of a feline (did I mention people love cats?). But what if I told you I meant Cat as in Caterpillar industrial Vehicles, or maybe I'm really talking about Cat Deeley the TV presenter (who has a different cultural context depending on what side of the Atlantic you are). Or maybe the Town in Turkey, and the list goes on.

So that's discourse in a media sense. Semiotics is the reading of signs and symbols, but discourse is using one thing to mean another if you like, or least the understanding that what it means to one person, isn't the same as the next person who sees it. So back to the E4 video.

First they considered the targeted audience, what do they watch and who do they look up to, their influencers. The most important part here is it's someone they relate to, someone they know and someone just like them.

Then there's the language they use, again, it's how they speak and it's very heavy on innuendo. You're not quite sure what they are talking about at first, but you're intrigued and want to know.

This relates again to not thinking with your head as per Randy Olson.

We talked about how you need to get out of your head and use your heart, well there are two other places as well. There's your gut and a bit lower still, your sex organs.

Here the language they use is more primal, it's sexual, and this is a language everyone understands (seriously, read his book as soon as you've finished this). So they researched their audience, tailored the language used for them, and turned voting into a primal action.

So lets turn this into your #SciComm video. We'll assume you've researched your audience and have a subject, but so far haven't written a single word down for the script. Good. Keep it that way for now.

Here's the thing with video and I'm going to take you now to William Goldman, the writer of *Butch Cassidy and the Sundance Kid*.

He's written two books on the screen trade and

screenwriting, but specifically I'm talking about *Which Lie Did I Tell?: More Adventures In The Screen Trade*, there's a couple of very relevant things he says that you need to know. The first is he says in Hollywood, 'nobody knows anything', meaning no matter how much research they do, they have no idea how a film will perform before it has bums in seats that have paid to be there.

This is important.

If Hollywood doesn't know if a film will be success, why should you think your carefully crafted video will be? Those of us old enough to remember before *Titanic* came out will remember all the rumours that it was an over budget disaster.

Then a film blogger was shown a preview, and he said it's a great film you must see it, and people started to think differently. They held the official pre-screening and they loved it! Suddenly there was a different buzz about the film, word of mouth was building.

But according to Goldman, he was talking to a studio exec a week before it was released, with all this buzz building, who crossed his fingers and said he hoped it would at least make $100 million as then the studio would be ok.

It was the first film to reach the two billion mark.

So you have no idea how people will love your video, all you can do is work hard to make is as great as it can be.

And the way you do that is think like a filmmaker. Randy Olson, a scientist turned filmmaker points out that film is a visual media, you watch it, don't read it.

But Goldman breaks this down to something even simpler. "Show, don't tell". What this means is people have a high media literacy, they understand the language of the medium without realising it. We all know when the big surprise in a movie is coming, we know the signs. We can tell who is having an affair with who at a dinner party as we can see the signs without a word being spoken.

As you watch TV or a film, you are unconsciously reading everything on the screen at 24/25 frames a second.

There's a wonderful term, Mise-en-scène, which means the entire image in front of you, so you're interpreting the scenery, the wardrobes of the characters, even the small props they are using. All of this information builds up an image telling you what is happening, or giving details of who they people in it are.

Everything on the screen tells you a story.

An easy way to explain this is to look at cooking shows. First of you can tell it's a cooking show as they are in a kitchen, but what sort of kitchen are they in? Are they in a restaurant or is it a traditional cottage style kitchen, a sleek all white one that's very modern looking, or somewhere in between?

Everything you're seeing is telling you something about both the chef and also what type of food they are going to be cooking.

Then how do they show you the cooking, first they talk about and show you the ingredients, and often they will show you what it will look like. As they go through the process of cooking, you'll see the shots in a repeated format. Wide, medium and close ups, each of which will show the cooking process and the chef.

When they do something different in the filming style, the show is always a hit, but they will rarely step far outside of this format, it's just the extra bits added that make you notice there's something new. Just like when giving a talk and the audience gets bored so you change a small thing to make the whole seem more interesting.

Cooking shows are a great example of 'show don't tell' as I've never seen anyone follow a recipe direct from the TV. To be honest I'd love to see that in real time.

Although they are telling you what they are doing, you're too busy watching it to listen. They are showing you in detail, but only telling you once. This is because what matters in cooking shows is the end product, not the process. We care about how it looks and how we imagine it will taste, not how it is made.

It's all about the what, not the why. People take away small bits from these shows that they may try themselves. It's the concept not the details.

The format of a cooking show is a good pro forma format to work with for how to do #SciComm but as with everything, take the format and change it. There's no reason why you have to follow it, but you can learn from it.

It's important to remember that cooking shows are about the what, not the why, and that science communication is about the why, not the what.

This is my biggest criticism of a lot science videos I see online. If you show an experiment, I can see what happened. What I'm interested in is the *why* it happened. You don't need to explain the *what* to me again. I know when the cake goes in the oven what will happen at the end.

Show me what happens/tell me why it happened. This is the rule of science communication videos.

When I see a video where it's constantly *tell* with little *show* I get a little heartbroken at another lost opportunity. This is really annoying to filmmakers (although I worked in social media marketing, I'm a writer/filmmaker at heart). Vlogs are great, but mix it up a bit to make it interesting. The camera doesn't have to be locked in one place.

So lets apply the show me what happens/tell me

why it happened rule I've just created for you to follow.

Now when it comes to working out your script, don't write text for what has happened, only explain it. You can show it again, and in close up while you explain it, this reinforces to the viewer what you are talking about.

This is another tip from William Goldman: audiences love 'how to'. It's why *Ocean's 11* is such a popular film.

So science films that explain how it happens can be related to this film. You're shown how they really did the heist at the end, but they don't explain it to you, they guide you in working it out.

Show me what happens/tell me why it happened. Now you have an idea how to write what you want to say, worked around whatever you're going to show. Once you've written it, put the pen down, get a stopwatch and read it imagining every step while timing it.

However long it is, it's too long.

It's easy to say that you're not making this video for TV or Film, you'd be working with filmmakers who would help and advise you and constantly rewrite your script. You're either making this for online or to be used for a demonstration or talk.

Think about how long you watch a video online for and at what point you get bored and start looking for something else. Most people will only watch the first few seconds, if you've lost their attention……..

So how long should it be?

Aim for around a minute, that's short enough that people are more willing to stay till the end, but long enough for you to say and show what you need to. And more importantly it will show a lot more than you think.

You can make them longer, there's nothing wrong with them being longer, but lets start with shorter ones, it forces you to sharpen up your storytelling. There's a great video on YouTube by *Physics Girl*, with the plate making two vortexes in the swimming pool.

Well the second half of the video is about the refraction of light and why it makes the shadows of the vortex.

Personally I would have made that as a follow up video, because there are now two things to remember in the video, not one. The first half is the vortex and the second half is the shadow. It would also obviously cut down on the length.

This leads into a useful tool on YouTube that I'm sure you know about already. You can see the viewer retention letting you know exactly where the viewer got bored and looked for something else to watch. There's a lot of noise out there to distract you and YouTube has all of it. Reviewing these statistics will help you become a better filmmaker.

You have to make a lot of videos before you get good at it, something Robert Rodriguez talks about in his Ten Minute Film Schools. He said that he expects everyone's first three films to be rubbish as here you're making mistakes on the technical side (hands up all those who haven't turned the mic on during an interview), and also learning the rules that go with making a video. Never forget that Richard Donner used to direct *The Banana Splits* kids TV show.

We all have to start somewhere so use your early films to experiment to find your style and don't be upset when it's not what you expected. One day, I may show you my first filmed news story. The cringe factor is easily set to eleven!

This section isn't a going to be a technical explanation on making your video, there's too much to go into. I would recommend you have a look at the film school videos on Vimeo, there's a lot of good tips and ideas on there for you. The good news about making videos is that it is really quite simple. Your phone can record video, you can probably edit on it as well and the quality is easily good enough for most of your needs.

You can get video editing software easily enough and Windows Movie Maker is a lot better than it used to be, although you'll be limited on the creative side. When I first started editing I used Final Cut Pro 7 (FCP 7), a professional editing programme and it looks confusing at first. In fact I really didn't like it.

Once I got the basics, how to add videos to the timeline, and then how to do transitions from one shot to the next, it started to make sense. As with all of the editing software, the more you use it the more you'll learn about it. And if you can use one of them, you can use all of them as they all do basically the same thing.

FCP 7 has been replaced with FCPX, which like it's predecessor you'll need a Mac to use, but it is both simple and powerful to use. And even better, you can still work on your edit while it is rendering.

HD cameras are cheap these days, more than enough for what you need (but get a camera that you can plug in an external mic). Viewers will excuse poor video quality, but not poor audio.

Graphics are easy to create, a lot easier than you might think these days, and you always be able to find a student who can make them and will be happy to help for the experience (or a pizza). FCPX has a lot of pre-set options that look very professional and even Windows Movie Maker has some but again you are limited in your options. For you time is a big consideration, so it may be easer to go to the media department and ask for advice, they'll find you someone.

You can even ask them to help you make the video, giving you a chance to learn some of the rules of filmmaking and the technical side of editing. People in the media departments are always going to be happy helping you, and media students more so.

One other thing to think about is of course music. Are you gong to have a jingle or something just playing through in the background on a loop? The biggest thing to remember about music is it is easy to add, but much harder to remove once it is there.

I made a short film with some drama students, just for fun and it was all improvised. They created the story and I filmed and edited it. When it came to adding music, I used songs already out there, and knew this would be an issue with adding it in.

The twenty-minute film has songs pretty much all the way through, but I wasn't able to remove it unless it added to the story to build suspense. When ever I watch it now, I regret that decision, but it was just a fun film, not intending for it to be shown.

Adding music does help your video, but you will need to consider how you do it, and if it will detract from what you are trying to say.

Plus of course you need to consider copyright issues.

Lets talk about whether you need a presenter or not. There are plenty of videos that only have a voiceover which are great, but unless you have someone who can do really entertaining graphics, you'll probably want to avoid that unless whatever you're showing is really entertaining.

So you're going to have quite probably appear on camera. You could just do just all close ups of the action with a voiceover, but people like to know who's talking, so yes, you will have to appear on camera.

What style are you going to opt for? You could go for the classic TV presenter style, or you could go for a more vlogger style. Both of these of course will need a script and here is a good tip. Don't memorise the script.

You're not a TV presenter or an actor, you're a scientist so trying to remember it is just a waste of time. Sure you could get a teleprompter app, you can get free ones that work very well, but it's much easier to write a script and remember it as bullet points.

Use the editing to hide all the stops and starts, make sure you shot enough cutaways and filler shots of whatever it is you're trying to communicate to hide the cuts. It's how it's done and is easy so long as you have enough stuff you can use to hide them. Or you can slow down the video of the action, or zoom in on it,

The videos I made for British Science Week that started this whole journey for me featured people from work who needed to be coached through the whole process. As we'd worked together getting the script right I made it very clear to them that they mustn't try and remember anything on it.

On the day I explained the set up to them, using two cameras and how I wanted them to sit, not on a chair, but perched on the back of a chair, so it rooted them to the spot, but they were relaxed enough to not be standing.

I'd set aside an hour to work with them to record a one minute thirty second thirty video. The room was set up with the cameras and the lights were turned on in the room. I didn't use any other lighting or reflectors as digital cameras are good enough to mean I didn't need to worry for these purposes.

They were away from the wall to avoid any shadows and I'd done some test shots using myself to see what places gave the best light. I could have used a light reflector as well, but I'm still waiting for CRISPR gene editing to give me another arm.

(CRSIPR was the lead story on BBC news today, and I wonder how many will think it can give you a new arm?)

The set up was simple and you can see how I'd designed it in the image below, and the presenters pretty much had the cameras in their faces. You'll also notice they weren't in the middle of the shot as this makes the viewer feel uncomfortable. I'd purposely left some extra space to include graphics and to consider the rule of thirds. This is a simple technique where you imagine a grid on that splits the screen into thirds, both up and down, and you position whatever you're filming over them. Again, these are rules that can be broken, but visually it is more appealing.

The second camera was there to allow me to hide editing cuts to smooth it out and look smoother.

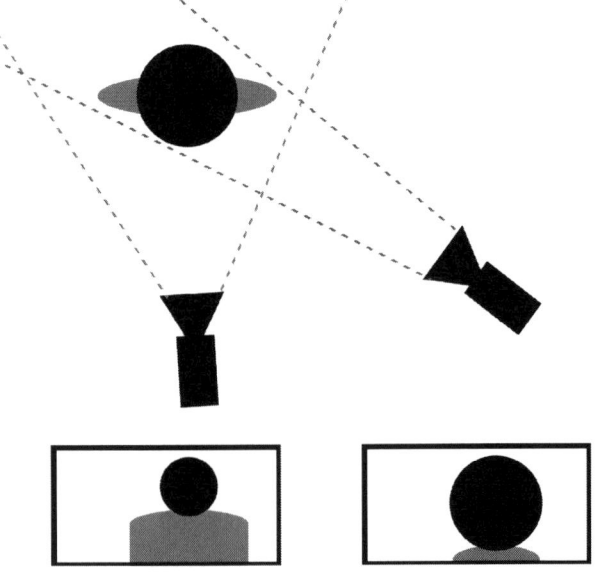

I could have used the vlog style hyper edits, but

decided against this, I wanted it to look and appear a bit more professional as I needed to get across that the person speaking was an expert in this subject. Hyper cuts are great for making it look like someone has a lot of energy, but really it's to hide that they keep stopping and starting.

And that's why I didn't want them to learn the script. They'd been involved with the writing of it and knew what the concept was, not the details of each word.

What I had them do was read a sentence or two, turn the camera on and had them talk to camera. Then I would get them to do it again, guiding them into doing it each time as if it was the first. Once I was happy, we moved on to the next couple of sentences.

Because I had planned each video down to the second, it meant I also knew what graphics would be displayed on the screen, and also when they wouldn't be on it.

Again, when I knew they wouldn't be on the screen, they weren't allowed to read from the script as they needed to sound natural. This is the hardest part for anyone to learn, and with my presenters the same thing happened for each of them.

At the start they were multiple takes for each sentence, but as we got further along and they relaxed into the role of being a presenter, they got better and better. Not learning the script allowed them to act natural in front of the camera and while talking to their own reflection in the lens. Here what you need to do is pretend the lens is a person you are talking to, just like a radio presenter imagining the clock is a person.

Acting natural on camera isn't easy and takes practice to do so. When your early ones don't quite work out, don't worry, put it down to experience and learn from your mistakes to make the next one better.

Get some feedback, it really helps.

One of the best bits of feedback I got early on was to do with a video I made for a friends martial arts club. In it he was holding a katana and I did a close up going a along the blade.

When I see it now it's one of those cringe inducing moments, as the blade looks static and dead. The feedback I got from my lecturer who was a former TV news producer was simple; get them to talk about it as they hold it.

What happens then is the object comes alive on the screen, it's no longer an object in the hand, it's a living thing. It works for everything, even apples and test tubes and anything someone is holding when you do a close up. It's such a simple tip and it keeps your video filled with energy, rather than suddenly falling flat while you show an object.

And that's the whole point of what you're trying to do in science communication, you want it to come alive for the viewer, not just be sitting there on the screen like a dead thing.

The hook you are using to get them watching is the science you want to show, but as soon as anything goes a bit flat they can become distracted by all that other noise and start looking for something else to watch.

What will keep someone watching is the story you are telling and this could be a difficulty for you. The hook is the thing that gets them interested in possibly watching the video. But it's the story that keeps them there.

It doesn't matter how great your graphics, camera work, editing or presenter is, you have to tell a story from start to finish.

Now we know the story doesn't have to be in a linear style, it can be non-linear. If you think of the quantum world and how weird it is (apologies to Jim Al-Khaili, I still have difficulties with the basics), so the story can be in any of the 16 states (¯_(ツ)_/¯).

You can start with the effect and then explain the cause, or follow the path of the particle and jump from one place to the next instantaneously, so long as whatever you do contributes to the story telling.

But as with Schramm in the communication models, following the fields of experience for communication there are some styles of storytelling that have been around as long as people have sat around fires.

The heroes' journey is a good starting point and a quick search online will bring up many versions of it as well as scholarly articles. Pixar also has one online which may help you as well, but the biggest point with both of them is that the stories are about people and their emotions.

In science emotion has quite rightly been removed, but that doesn't mean that the scientist can't display it. Maybe the story is about how you discovered the science, or how someone else did. Now you know how to tell the story as the story has either been written already, so you just need to rewrite it for your audience, or tell your own story.

Your story of course is much better than someone else's, sure the story of Fleming discovering antibiotics is great, but what about your story for how you are applying that cultural history into your research to help others.

Finding the story is probably the hardest part and will require you to change the story as you create it. It may be that the hook you first started with changes as you develop the story as you realise there's a much better one.

Storytelling requires you to be flexible. When I took my first steps into the media world, it was writing scripts on spec. I never sold any of them, but I got a lot of great feedback for them. When I was writing I would also start with a scene in my head and go from there, seeing where the story took me.

I would rarely plan ahead what I was going to write as so many others do. Maybe Shakespeare had a plan at the start for Hamlet, but personally I think he started with someone seeing a ghost and his genius did the rest (and yes, it was him that wrote it!*).

How you create the story you want to tell on video will change as you get better at making them. But again, as Robert Rodriguez pointed out and my own efforts prove, you'll suck at it at first, but you will get better. Don't let any bad early videos put you off, you don't have to post them, just get that feedback to get better at making them and your own voice will develop.

*Cultural reference to *Morecombe and Wise*, look them up on YouTube.

"Social media is not about the exploitation of technology but service to community."
Simon Mainwaring

4.4 How to present your message on social media

I started working in the social media communications while I was still at university, the first time round. I'd made the switch in career and was doing a Foundation Degree at Highbury College in Portsmouth in Professional Media Production.

This was mostly a journalism based course and was more on the practical side than the theoretical. We were making videos, writing newspaper and magazine articles, TV and radio shows, all hands on doing it as if it was for real. It was a pretty intense two years but the skills I gained are still serving me well.

There's a reason Highbury College is well respected for training journalists in the UK, and that's because they work you hard to produce professional quality work from very early on. The training was good enough that six weeks after starting the course I had my article published in the local paper.

It also gave me a career, something I can't thank them for enough.

As the end of the first year approached, I was asked to help the marketing department with some video editing. About two days later I was asked if I could work there over the summer filling in as the Digital Media Officer, and of course I said yes.

Most of my work was video based, but some of it was also using social media, something I hadn't really given much thought to (we doing were digital in the second year).   Besides, it's really easy isn't it? When I started there I didn't even know how to log into a Facebook Page.

I must have done something right as they asked me to work for them again the next summer after my foundation degree had finished. By this point I was sick of studying, I just wanted to get out there and work, so I went back to my crappy call centre job and every now and then looked for a media job.

Then 42 minutes before applications closed, I found a social media job with an emergency service and I applied.   But I wasn't interested in the social media part, it was writing and editing the internal magazine that interested me more.   I didn't think much about it really because everyone working in marketing obviously knows how to use social media, so I never gave that side of the job much thought.

Boy, was I wrong.

The very first thing I was asked to do at the job was write a couple of pages explaining social media to my colleagues and then present it at the weekly team meeting on Thursday.

Have you ever heard of Imposter Syndrome? I had it big time right then.   All I kept thinking about was how little I really knew about the subject and how they would catch me out in this very obvious hole of my knowledge.

So I delivered the paper/presentation, feeling my ears burning and I realised that they knew nothing about social media marketing and communications. And that's because just like in science, people specialise in communications.   The little I had done for the college was more than a lot of marketers had ever done in their careers.

Now more new graduates are coming into marketing, more and more marketers are better at social media.   As Stephen Waddington, the former president of the Chartered Institute of Public

Relations points out, PR and marketing requires a constant upgrade of your skills to be effective at it.

So to jump forward a few months and I'd been ask to deliver presentations to all sorts of people on how I saw social media communications, and my guidance notes were used by many others (and became the basis for the first draft of the Government Communications Service Social Media Playbook).

All I was doing was applying my knowledge of the media gained through practical experience into the social media realm. Literally all I ever did was to pick which of the many great stories generated across the agency new and look for ways to tell people about them.

I learnt something important very early on while reading the various social media blogs looking for new tips, and trying to understand the various algorithms to help the reach of the content. But it seemed like a blasphemous thing to say and I was afraid to discuss it with anyone. I had written a blog about it, but hadn't shared it.

No one wants to be the first to say that the emperor is naked.

Maggie, the head of our communications department used to call me a social media guru, although I preferred Tsar as it sounds a lot cooler. We both went to a talk organised by the industry that was held at the university I have now just graduated from and about to do my MA at. We didn't learn anything new there, the talk was at a more introduction level to social media marketing and we were ahead in this game.

Most of the talks that night didn't impress me, apart from one, and I sought them out to say how much I agreed with the most important thing he said.

"There are no experts in social media".

This was the subject of a personal blog I'd written a few weeks before, and the reasoning is, although the skills I have in social media communications are transferable, the experience isn't. What that means is what works for one campaign, won't work for another, even in the same subject area. This is what the other speaker said as well.

I've since seen this a few times now, and if you ever hear someone say they are a social media expert, know this one truth: They aren't.

So if you're hoping to find something in the following pages giving you concrete advice on how to 'do' social media, looking for something from an 'expert', you won't find it here. The advice I'm going to give is good advice based on my experience, knowledge and skills to help find your own way in social media.

Before I start with the advice and going through various parts of the social media world, I want to begin with the term I used to explain to government communicators how to talk to people on these platforms.

When explaining how to phrase a message on social media, I would describe it as 'officially unofficial'. What I meant here is that it's from a government agency, so therefore everything you say is official and is likely to be quoted by journalists, so you'd better be careful.

But you're also on social media, and the key word for this type of media is 'social', so your official message shouldn't sound like it's official, you need it to sound unofficial. Finding this balance isn't easy.

Being official relates to you as scientist as you have to be factually correct, but you need to balance this with the unofficial bit. Think of it this way, $F=ma$ is factual and official, but it's also boring.

So how about explaining it the same way as if

you're pushing a car? Most adults at some point in their life will have a car that breaks down, so here they can instantly see the relationship of the equation. You could even remove the wheels and start talking about friction as the next step.

All I've done here is remove the equation from the picture and replaced it with something that people can instantly relate to. The official to the unofficial.

As Randy Olson says, 'don't be such a scientist', get out of your head (official) and into your lower organs (unofficial). You have to use your emotions to step away from the message.

Think about what you're trying to say, not what you have to say. Being too 'official' turns people off and this damages your organic reach.

Your organic reach is how many people see what you post without you paying to promote it, and sadly it's a number that is going down. Everyone will tell that this organic reach is down and the reason for it is obvious. Social platforms want to make money, and they do this by restricting your reach, which encourages you to pay to promote or sponsor it.

The good news if you do decide to do that is it's quite cheap. The bad news is if you do it a lot you look cheap. Use it sparingly.

Something of course you'll need to consider is when things go wrong. Maybe someone posts something to the wrong social media account, or they act in haste in responding to someone.

Well, they won't be the first, and they won't be the last. I'd recommend having a look online for social media fails, there are many and they are quite often both funny and shocking at the same time.

One that I used to use as an example was when Police Scotland trolled an author who was on a political TV show at the time. If the author hadn't defended the mistake of not checking which account the tweet was sent from, they would have lost their job.

Many others have lost their jobs for doing a lot less than that, but my view on this is to relax. Accept you will make a mistake. Worry about how you handle it, and have a plan for doing so.

It will make your life a lot easier in the long run.

*"Words, words, mere words,
no matter from the heart."
Shakespeare, Troilus and Cressida.*

4.5 Writing the message for social Media

When writing your message for social media, it's important to remember Lord Reith and how he summarised the British Broadcasting Corporation's role, 'to educate, entertain and inform'.

For social media the order is 'Entertain, inform, (and if you're lucky), educate'. The last one will hopefully happen as a by-product of the first two, but it shouldn't be the primary reason for you use it as a channel to communicate.

Social media is used as a distraction for most people; it's the thing they do while travelling, watching TV, eating, and the list goes on. People reach for their mobile device to look at social media when they are a little bit bored.

Other than YouTube, people are using it to give themselves something to do for a few seconds (YouTube of course is something you go to for a reason). You must remember this at all times when posting anything to social media. Your purpose is to be a distraction and give them something that makes them want to stay a bit longer.

So all the earlier bits here about how to create and structure your messages, this is where it really matters. Here you are combining every method into one, using all the communication models we've looked at together.

And of course, you're in a constant feedback loop allowing you to refine and improve your messaging all the time, look at your stats and metrics to work out what works.

But here's the thing, writing for social media is

very different to writing a feature or a blog. People are scrolling quickly and stop when something grabs their eye and you must grab their eye to get some engagement.

Remember those algorithms from earlier? Without engagement you won't get them doing what you need them to. For people to see what you post, you need people to engage with you. This means you have to stand out and you have a fraction of a second to do that.

The text you write has to engage with them, and it has to engage on an emotional level. So Randy Olson talked about the four centres: Head, Heart, Gut, Sex organs.

Unless you're trying to communicate biology, you'll probably want to avoid the last one. It's not just an ethical nightmare but as Randy points out, this is the one we have little control over. Blame the biologists.

The head lacks any emotions that will make someone stop scrolling, the others will.

And that isn't about what makes you stop scrolling, it's about what makes your audience stop scrolling.

So lets stick to the 140 characters of Twitter for this section, so long as you promise to remember that the text you use on one platform can't always be used on the others. On Twitter people use a form of shorthand that in most cases won't make sense on Facebook.

Let's stick to the $F=ma$ example. F stands for Force, m for Mass and a for Acceleration (please remember I'm not a scientist, so if I make a mistake it's in my understanding of it).

So the force required to move the car is equal to the mass and acceleration. In this example we're not using any pictures, videos or gifs, just 140 characters.

Oh, and you're doing this as well.

The first thing I'm thinking about for this message is how much easier it would be if we used a few emoiji's but I'm being mean, I want you to use words.

"The force needed to push a broken down car is equal to the mass and acceleration. F=ma"
86 characters.

Falling asleep yet? How about;

"Pushing a car uses F=ma, Force to get the car (mass) moving (acceleration)"
74 characters.

A little better but still quite boring. The first example is very much in the official camp, and the second starts to include the unofficial part, as it sounds (almost) like someone talking.

You might have noticed I didn't say the car was broken down and wondering why. Well ask yourself this. Why would you push a working car?

One of the best examples of sub editing relates more to twitter than anywhere else. My news writing lecturer, Karen Nicholson, got us to sub edit down a four word sign outside a shop, to remove all unnecessary words.

FRESH
FISH
SOLD
HERE

So what words do you think can be removed? I'm not answering yet. The idea here is to see how much you can cut out that isn't needed, and these words I'm writing here aren't needed, I'm just filling in to make you go back and think about the question.

Seriously, I'm not going to tell you for a few pages yet so have a think about it.

Sub-editing is a brutal process and it takes practice to understand the process fully, and to get the sign down to one word.

When I did this, I thought it was two words and I was genuinely surprised it could be reduced to only one word. I think I've teased you enough now, so we'll go through it one word at a time.

The sign is outside a shop, so we don't need the here. It's a shop so what else are they doing than selling things. And of course it is fresh as no one buys rotten fish.

The only word you need is 'fish'.

How many did you get it down to?

So now you can see the process, I'm hoping you can understand why the car that is being pushed doesn't need explaining that it's broken down. The audience is smart enough to work it out.

So back to the pushing the car Tweet. To be honest, I can't think of any other way to write that in the limited format of twitter, but I'm interested in knowing if you come up with anything better using just text alone.

And that's the limitation of just using text, you only have a fraction of a second to get their attention and break into the bubble created by their social media algorithm. And if you can't get in the club, you won't get to have a dance.

Text only posts on social media are the equivalent of wearing the wrong type of shoes when trying to get in a club, and the algorithm is the bouncer not letting you in.

You need something else to get their attention.

"Which of my photographs is my favourite? The one I'm going to take tomorrow."
Imogen Cunningham

4.6 Pictures – a Jpeg says a thousand words

It's a cliché, but the thing about clichés is they are often true, this one more than many others. An image changes whatever you post and people will always look at the image before they read the words. Even on Twitter if they've turned off image previews, they are still more likely to want to look at the image before they read the words.

This is both a good and bad thing. An image is a snapshot of time, and often lacks context. That's why you see more and more, text on the image. And here's a stranger thing, text on an image is more likely to be read than the text that goes with it. I'm going to come back to this point later, but for now, let's focus on the image.

When looking at a website, the path the eye follows is in an F shape. Left to right, down a bit, left to right, down a bit. For photos this tends to be more of a Z shape, but if there's a face anywhere on the image, the gaze will stop right there, even for a fraction of a second.

Our brains as I'm sure you're aware, are great at identifying faces, or even the shape of a face and this is nothing new. The Makapansgat Pebble, dating from around 3,000,000 years ago attests to this. It was found by Australopithecus and carried back to their cave, some distance from any possible natural source for making a smooth stone with what looks like a face on it.

These patterns can be useful to you, guiding the viewer to where you want them to look. There's been a lot of work done on this by many scholars and researchers, but as with storytelling, there are rules you can break. As with filming, you can break those rules to jar your audience into an emotional response.

Images are the same, you just need to understand your way around them. So lets start with the basics. What are you trying to say with your image? Are you trying to support the text, tease the audience into reading the text, tell a story or just show something interesting?

If you're trying to support the text you've written, and want a photo to go with it, it has to be relevant. If you are saying you're looking at something with a telescope that night, what sort of image would you use?

Maybe a telescope from outside, telescope from inside, someone looking a screen, the thing you want to look at, the night sky or an image of a celestial object.

That's just me thinking of where I want to start, and I could use any of them. What this should tell you is you need a photo library, a wide selection of random images that you can use, and then reuse (after a period of time), over and over again.

I had over 400 images I collected to use for the Coastguard and I was constantly adding to them. They were in folders based on what was in the photo as well as general use ones that didn't fit any of the normal headings.

I'd be very surprised if you don't have a smart phone, but here you always need to be thinking of taking photos that you can use at a later date. You could use a commercial photo library, which are surprisingly cheap in most cases, but I'm guessing you don't have a budget, and any photos taken by you have a few advantages.

First, you own the copyright, so the image is free

(but if you take it in the line of your work, it belongs to your organisation, so check local copyright laws). The image is also free to use and you won't see someone else using it.

Imagine the embarrassment of using the same photo that a politician is using! Or even worse, someone could be using the same photo to represent the very opposite of what you are trying to say.

The biggest advantage by far is the image will always be relevant if it comes out of your own image library, because it directly relates to you. It doesn't matter if the picture doesn't quite match up; here what you are using it for is to convey a 'feeling' for what you're saying.

This comes back round to the emotional part of the message, conveying a feeling behind what you're trying to say. We know that most of communication is non verbal, and I need you to think of the written part as verbal communication with no subtext.

We all know that it's easy to apply your own emotions to an email you've received, so you do everything you can to avoid any misunderstanding when sending them. That's what you must always try to avoid, and a picture can help with that.

One thing you should always include in photos as much as possible is a person. They don't have to be smiling, but they should look approachable. And never use a smiling photo where the story is potentially a sad one or has negative connotations. The same goes for humour in the photo that you might miss. This would be where you create that discordance in the audience where you don't want it to happen. Check your photos carefully to avoid any dumb mistakes, it can happen so easily.

I did this myself once by accident during an exercise to simulate multiple people in the water. The photo was taken by a Coastguard in the ops room preparing the simulated casualties that were going in the sea. They were writing numbers on oranges that would later be rescued and it was good way to demonstrate the preparation that goes behind the work they do.

Out of the 100 numbered oranges, I didn't notice the number on the one they were holding up to the camera. Thankfully, not that many noticed the number between 68 and 70, but I should have. Always consider everything in the photo, it's easy to offend by accident.

When you're using a person in the photo, have them do something, do not have them pose for it. Or if they do pose for it, at least have them look natural and relaxed (maybe holding an orange with 69 on it). It just looks dull and too much like a stock photo or a photo call. People have high media literacy and can read an image without automatically and will see that there is something wrong.

This is the reason why you'll see press photographers blast away bursts of photos during press conferences. They are looking and waiting for the moment when the subject look normal. Whenever I had to take a photo of my old CEO, I would always take one just before he said he was ready. When he posed for the photo, he always looked like he was grimacing, and the picture always looked forced. He still wanted the posed photo, but I would only have this as a spare just in case.

One of the things I've seen a lot over the past couple of years on the social media blogs is the use of colour as a way of creating branding in an image. Now this is great advice if you're a company that has colour attached to it. Think of Richard Branson's Virgin (red), McDonalds (red, yellow and now green), Shell (red and yellow) KFC (red and white), Apple (white, but used to be multi-coloured) and so on.

Now if you're a global or national brand there's value in that. Each image should feature the colour that is culturally associated with them. But you don't have that, although it could be argued that in science, white is the colour people connect to it via the lab coat.

The only reason I've included this here is to stop you trying to think of a colour you can brand your images with. It's not going to help you if you see that advice and try to use it somehow.

There is a lot of good advice out there you can access, but not all of it will be any use to you. Can you create a branding to your images through the use of colour? Yes. Will it help you communicate science to the public? No.

The most important part to take away from this section is to make the photo not only relevant to what you are talking about, but to make it about people in some way so that the audience can connect with it emotionally.

That's why people are following you.

"No film ever ends up exactly as you would like it to"
George Lucas

4.7 Video on social media – moving pictures

Video has a lot more power than a photo, but on social media that's a double-edged sword. In most cases people won't watch a video if they aren't connected to Wi-Fi or have a great data deal from their mobile network provider.

There are ways around this, but first I want to give an example of the power of video over the written word.

One of the problems for emergency services, in the UK at least, is people will often post something on a social media platform or call their friends before they call the emergency number, even if they are the ones in trouble.

Being a social media marketer, I pushed for a simple and quick video, 20 seconds long, then the maximum time for Twitter. But I lost that battle, and a blog was written. It went through many drafts to keep it clear and explain the importance of why people should call directly and the dangers of not doing it that way, and so on.

If I remember correctly, around 80 people actually clicked the link for the blog, and the reach was very low, a lot less than 10% of the current followers. A dismal result for something so important.

The message failed because it was too long, and not interesting enough for people to spend five minutes reading it. I again presented the case for a 20 second video, to be posted the following week on the same day.

I knew the message they wanted to send and

spoke to the controller about it, said what I wanted and went there to film it. The video could not be simpler, it was an intro shot setting the scene for the location showing the person about to talk doing what they do. This included a banner at the bottom reinforcing what they were about to say, simply, call the Coastguard and it lasted less than five seconds.

The next section was them talking to camera stressing how important it was to call the Coastguard themselves and not get someone else to do it. This lasted ten seconds.

The last five seconds was showing them doing what they do with the banner at the bottom again.

Sadly I can't remember the reach or views for the video, but I do know it was the highest viewed video that month by a long way. More importantly it reached out beyond our usual audience, breaking out from the organic reach we all want and need to achieve.

And this is a boring message, reminding people what they already know what they should do anyway. As a blog people reacted as if they were being told how to suck eggs. As a video they were able to relate directly to a person, and gain an emotional connection. It was almost came across as a plea.

For you as a communicator of science, emotion is extremely important. Without it I have no reason to watch your video. I often see passion for a subject, but not often the emotion that's needed to connect to the audience.

Poppy of 'Poppy Presents' on YouTube is a great example of this emotional connection. She has bags of passion for what she is doing and this comes across. Where the emotional side comes in is she is truly enjoying what she's doing and this comes across also.

Because she is emotionally connected to the

subject, this transfers to the viewer. The Coastguard making the video, well you can bet he cared about it (I know because he called me asking what could be done as it was happening too often). This emotional connection to the subject came across, even as he is trying to hide it and look professional.

You see this all the time in films, if you think about the film Avatar, when Jake first wakes up in the alien body, you can feel his enjoyment as he runs and you engage emotionally with the character through this enjoyment.

Another actor gives a perfect example of what I'm talking about while talking about his role in a film where science was important to his role. Jeremy Renner was on the Graham Norton Show on 14/10/2016 with Amy Adams promoting the film Arrival and the subject of science came up.

Graham wanted to know how they connected with science as actors, giving the audience a look to apologise for talking about science. Think about that for a second, what he was saying with that look is that science is boring, cold and not very interesting.

Amy Adams sort of agreed with him, saying that it all went over her head as she's not very good with science, but that Jeremy did understand it.

This is the important part Jeremy Renner said;

"I got an idea of what we were talking about, but it's not really the most interesting thing to talk about, so I had to create some emotional content to zero's and one's and binary, otherwise it's really quite boring. To be excited about zero's and one's I had to create emotional content to it."

There are two key things to take away from that little section of the show. The first is:

"…..not really the most interesting thing to talk

about…"

And….

"….it's really quite boring…"

This was reinforced by how Graham asked the audience to bare with them as they talked about science and the look he gave. Science is perceived to be dull.

And then we have the bit about how an actor approached the role of playing a scientist.

"….I had to create some emotional content…."

And…

"…To be excited about zero's and one's I had to create emotional content to it….."

A photo is a snapshot in time, but video is about time and space. You're capturing time and transferring it to another space, allowing people to gain a better understanding of who you are.

And you need emotional content to create a connection to the audience to capture their interest in what you do. Without emotion you're no better than Spock, who was played by Leonard Nimoy in such a way you knew exactly what he was feeling, conveying his emotion to the audience.

And here again we come back round to Randy Olson. To display emotion you need to get out of your head and into your heart and lower organs. Use your gut instincts to guide you, but only let your head be in control of the destination, not the route you take to get there.

The importance of this emotional connection can be seen by the difference between the blog and the video example I just gave you. The blog was all in the head, the video spoke to the heart.

But the blog was also a little condescending, telling people what they should already know allowing the audience to put their own emotional spin on it. The video removed that spin. And there's no easy way to say this, but the blog also had a speaking down to the audience feel to it.

In the blog it sounded like they were being told what to do, whereas the video it came across as being asked to do something. The blog sounded like the audience was stupid, the video like we needed the viewers help.

Which brings me to a huge problem in science communication. Talking down to the audience, and it can happen by accident and is an easy mistake to make.

Randy Olson gives an example of it happening in a climate change debate and how the scientists lost the audience with a simple turn of phrase. I've seen Neil deGrasse Tyson do it, but get away with it because the audience agreed with him.

He said, "Make America smart again."

Think about that, if he's saying it needs to be smart again, there's two things he is saying. It used to be smart, and isn't anymore. By using a Trump campaign saying, he's turning it into a targeted phrase towards both Trump and his supporters. Subtext is everything.

When you are talking with emotion, you're not talking to the head of the audience, you're speaking to their emotional centres, which then translate the message for their head. Every politician understands this, or at least the truly successful ones.

You might be asking yourself if you're able to use emotions in that way, well, "Yes we can."

Use who you are to inform people of what you do and why it matters to them. Consider the emotional content of your video to get the idea across, is where you are using Schramm and the fields of experience to find the middle ground needed for effective communication.

The audience are unlikely to have the same field of experience you have in science, but they do have the same emotions you have. Unless you really are Spock, then maybe communications isn't the best place for you.

Ask yourself what you are feeling and how you are connecting to the cold logic needed for science that is inspiring you to want to tell people about it. And then remember that at every step of making the video.

By this point I'm hoping you are now ready to start making your video and know how long it will likely be. As I said earlier, shorter videos work best, but if you have to make a longer one, then the emotional side is by far the most important.

Now if you're sharing a longer video on social media, you should automatically accept that it's unlikely to be watched unless you have this emotional connection.

Even then there's no guarantee they will watch all of it.

But this is the great thing about Facebook, Twitter and Instagram. You can upload video to them directly.

I can't stress enough how important this is and that you should always post the video onto the platform rather than sharing the link. Like all rules though, they can be broken, and the best time to break them is when it's a long video.

In that case think of YouTube as the cinema and the social media platform as the trailer. As soon as you've finished editing your video, make a trailer. Then you can post the longer link and the trailer becomes a teaser. Hopefully this will encourage them to click the link, or at least remember to go back and watch it another time.

You also need to consider the volume of your video, but in a way you may not be thinking about. Unless someone is wearing headphones, in public they will probably have the volume of their mobile device turned down.

On YouTube this is less of a problem, but maybe it's worth put subtitles on your video to help these people. While they may be able to save the video to watch later, there's also the chance they won't get round to it as the noise of their life gets in the way.

There is one other option for video on social media and that's broadcasting live. This gives you a lot of advantages. If you are at an event, you can let people see what's happening right then.

You can let people see directly into your world in an unedited fashion and the spontaneity of live broadcasting will help you connect emotionally. You won't have time to think with your head, you will only be using those other organs.

A journalist friend of mine told me a story about how the local Police used Facebook live to broadcast a statement following a court case. For journalists, especially on radio, this is pure gold. They can take the audio and use it air as if they had someone there.

But more than that, it's another example of using the media to compress time and space. Here the audience is being transported directly to the event to see what is happening right there.

Live broadcasting has both an intimacy and connection to who you are and what you do. But don't be like that police force who are scared of social media. They tried the live streaming as an experiment, but have decided the risks of doing it are too big to do it again.

For you live streaming from an event allows you to reach out beyond the immediate to those who can't be there.

It's another way of the media changing time and space to transport the audience faster than 12 parsecs to where you want them to be.

Jake: How often does that train go by?
Ellwood: So often you don't even notice it.
Blues Brothers

4.8 Frequency – Saying the right thing at the right time

This section will be a lot shorter than the others, but still very important. I've seen all sorts of advice from social media 'experts' about how often you should post content and the simple fact is, most of it is wrong.

Quite simply, you post when you have to and when you need to.

There is nothing worse than posting something because you haven't posted anything for a while. There's a well known TV presenter in the UK called Richard Madely who came into a lot of criticism a few years ago about tweeting he was drinking a coffee. He defended this by saying he was told he should tweet four times a day, every day and knew he needed to post something.

So here's the thing, the advice was both right and wrong at the same time. Yes, you need to tweet something, but what you tweet needs to be relevant. If you're going into a meeting, that's boring. But if the meeting is about something your audience would like to know, that's not boring.

The problem is knowing which is the right thing, and again, this comes back to knowing your audience. You have to give them what they want, not what you think they want.

This is what decides your frequency, how often are you doing something they want to know, versus what you might think is interesting. Sometimes, you might have too much to post on a day so that it could appear you're spamming them. Then you need to self edit.

Sometimes you'll be at a conference and sharing with people who aren't there, in which case you're not spamming, you're communicating.

Maybe some of it can wait for a day, a week or even a month. If you know you're making a video about the sex life of beetles in a month (sex sells), and today you're having a meeting about the research trip, maybe this can be stretched out over a longer period to give you a campaign.

You can see how NASA uses Instagram and Snapchat this way. They have a plan for what they are going to be posting, and then they have days where they do a lot of videos for Instagram stories and Snapchat. There's a power and immediacy to it you don't get if you are constantly posting the same type of thing all the time.

This is where frequency of postings becomes relevant. You need a long term plan for how you're going to post everything. I used to keep a calendar with all the events and relevant subjects coming up.

This allowed me to plan how and what would be posted and see if there were any clashes. Often I was only able to keep the plan in my head, and it would need a constant mental rewrite; that's the nature of an emergency service. But for you, knowing what's happening allows you to have some idea of how often you need to post.

The plan also allows you to identify any gaps, and this is one of the best things about having a photo library and also the videos you've already used. It allows you to reuse content (after a suitable time), to fill those gaps.

Throwback Thursdays are a blessing.

And of course the gaps are where you get to show a little more of that emotional context by talking about yourself, to let the audience in to who you are.

The other bit of advice is at what time should you post on social media. I've seen this a lot, but as I talked about earlier, organic reach is down, so timing must be important. Well, Buffer couldn't answer this, so don't expect me to be able to.

The best advice for when to post is when you want to. There are times like just before lunchtime, just before people go home from work and when they get up in the morning, but here you're trying to beat the algorithm and everyone else as well.

The more people who engage with your posts mean it's more likely to be seen by them, and who the algorithm thinks is like them. You could ask your followers to share the post, but frankly this makes you look a bit desperate.

You need to understand that very few postings will ever go viral, most will just be plodding along, giving you no idea if it's well received or not. The worse is when you think it's going to be loved, and no one cares.

I have too many examples of this happening to even try and pick one out as an example. That can be crushing, but that's the game. So long as you are posting content that engages on an emotional level, you're doing better than most.

But when it comes to going viral, it takes more than just posting it. To be truly viral, it takes traditional media as they have a much bigger reach than you ever will.

You may remember the double rainbow video and how it was everywhere for a while. Well that had been on YouTube for about six months before it went viral and hadn't gained that many views. What happened is it was retweeted by Jimmy Kimmel and then broadcast on his show.

Overnight it picked up millions of views, was downloaded by many and used on their own accounts and you know the rest.

I've seen first hand when something goes viral, and it's both exciting and a little scary how you lose control of it.

For me it was a video taken by a rescue helicopter of a sinking fishing boat. The crew had told us it was impressive to watch. Don't worry, everyone was rescued and no one was hurt.

The boat was taking on water and then turned over, dumping the crew in the sea and then sank. It took about 53 seconds to happen.

So the video was uploaded and within five minutes the press were on the phone asking for copies of the videos. Obviously they were all given a copy, but only about six were sent out to journalists, that's how few asked for it.

Over the space of the next hour I was watching what was happening on social media, tracking it as media outlets and individuals had downloaded the video and were using it.

Even landlocked countries were using it in Africa, South American and Asia were using it. The next day we were told it was shown on national TV news in the US by three networks, none of whom asked for a copy.

This only happened because as an emergency service journalists were following us. If you want anything to go viral, you need traditional media, either online, radio, or broadcast to pick it up.

Find your local journalists, and get them to follow you if you want to go viral one day. And remember, unless you have great content, don't expect it to happen.

"I think it's very important to have a feedback loop, where you're constantly thinking about what you've done and how you could be doing it better."
Elon Musk

4.9 Feedback – learning from your mistakes

I can't stress how important feedback is, and you might not always be able to get the feedback you want. Sure, you might get a few comments and replies, but you also might not. Don't let this upset you.

What you need to look at is the stats. You can keep a spread sheet of them if you like, to track them over time to try and find patterns, but having stared at months worth of them for hours I can tell you that you'll be lucky to find anything I haven't already talked about.

What they will show you though is if there is a particular subject that does better than others. This is a good bit of feedback, as you can then use the algorithm to help get the less liked subjects a bit of attention. But only occasionally, you don't want to over do it.

Lets say your sex life of beetles subject is really popular, and your little posts about the making of the video are well received. Well, this can be a good time to talk about a subject less liked, as the algorithm will help promote the post to those who liked the beetles.

This is feedback working for you. You've used the stats and metrics to find a problem and then found a way to overcome it. It might even be that you decide to follow up beetles with one about scorpions, but because one worked, doesn't mean the other will.

Just like the movies, no one knows if it will be a success. Maybe more people like beetles than scorpions.

You have to be careful about doing this type of thing as if you do it too often and people don't engage with the scorpions, it can impact what happens in the future to engagement for the beetles.

It's a difficult game to play, trying to use one thing to help promote another in this way. Sometimes it will work, sometimes it won't. As with everything in life, if you don't try it you'll never know. As Robert Allen said, "There is no failure, only feedback."

The other type of feedback is just like the one I talked about earlier when you ask people, except this time, you're asking those who follow you. Send them a message and ask what they liked about the beetles video, and what would they like to see next.

Surveys of course are a great way to elicit feedback directly and Twitter lets you do a poll for your followers so you can ask them what they want to see more of. Just don't be upset if the audience wants to see more of the one thing you don't really want to talk about.

But if you are asking the audience via direct messaging what they want to see, try not to do this too often. And try as much as possible to only ask those who have engaged with your posts regularly.

The audience is king and you need them, but if you annoy them, they can just as easily go away.

4.10 Trolls – feedback with an attitude!

Anti-vaxxers, intelligent designers and creationists. All have a history of trolling, although these days it could be anyone. Trolls should be looked at with a sense of fun, but that doesn't mean you should take the piss out of them.

There are two types of trolls, and how you respond to them will depend on which type they are. For both types though it's important not to talk to them as if they are stupid. It is highly unlikely they've had the benefit of your education. But saying that, I'm sure you've seen Katie Mack's masterful response on Twitter to the one, when told she should learn some actual science and *"stop listening to the criminals pushing the #GlobalWarming SCAM!"*.

Her reply was; "*I dunno, man, I already went and got a PhD in astrophysics. Seems like more than that would be overkill at this point."*

Now that was a great response, but was he just trolling or was it mansplaining? Looking at his Twitter feed, he wasn't trolling, it looks like he genuinely believes he knows more than a scientist. The key thing here is I looked at his feed, and I'm sure Dr Mack did as well.

So not all trolls are really trolls, just people with deeply held beliefs.

As a scientist you are used to arguing with people who believe something, presenting them with evidence based logic to support your position. This is useless against most people, and that is because deeply held beliefs are in the emotional centres, not the head, where we like to keep the logic.

We've all been angry and all know that this emotion overrules pretty much everything else; it becomes your primary focus, until you calm down. What people with deeply held beliefs need are calm emotional responses just like Dr Mack's.

Based on his reply about the need to get a refund on her qualifications, he was angry. And lets face it, he had been embarrassed by it. More so of course as this type of troll speaks from the emotional centres, and they don't speak the language of logic.

When the media picked up on the tweet, but not his response or the one from Dr Mack about how her PhD was funded, this was probably made worse and now his views are quite likely more entrenched.

It went viral thanks to the media, just like a double rainbow.

So the two types of trolls, which I teased you about just now, come in very different flavours. The first that I've just talked about come with a strong belief about something. The second are just trying to get a rise from you, the equivalent of Robert De Niro looking in a mirror and asking if you're talking to him. I want to talk about how to handle the first kind first, as it makes more sense.

Firstly, you're not going to win unless you can come up with a zinger as good as Dr Mack. As someone who ran a government social media account with a campaign group fighting against us, I have some experience of trolls (in a way I hope they read this).

At first I tried engaging with them, correcting bad information with the truth, fighting the battles that need to be won to stop bad information being shared. In military terms, I was trying to win the hearts and minds of both the public and the campaigners in our audience.

I made an effort to reach out to them to try and help them see the truth, but nothing worked. One particularly vocal troll of the group was literally seething in anger, all his comments towards us (often aimed directly at me), were dripping with rage.

I didn't always help reduce this.

There was one sentence on the agency's website that he seized upon, the length of the coastline of the UK. The webpage stated it was over 10,000 nautical miles, which is true; it is over that. The trouble is there are many different lengths recorded in many different places.

Hence the use of the word 'over'.

He really didn't like this, and went on and on, so I did a bit of research, and gave him the answers that I know you're all thinking about. Do you measure the coast at high or low tide, does a sandbank that's barely submerged at high tide count as an island?

All of this was put forward as a conversation in very different ways until they said 'Yes I know that." To which I would reply, "I'm glad you agree with us on how difficult it is to measure and why we say over 10k nautical miles."

This would generate a storm of comments and bile aimed at me. I'd talked them into agreeing with the very agency they hated with a passion. I always found it funny, but none of it helped.

The next time he raised it I then talked about how the coastline is fractally infinite, again he agreed, and the cycle happened again. The next time he raised it I was going to talk how the coastline was in a quantum state, you could know where it is, or know how long it is, but not both at the same time (not completely true, but good enough for my point), but they never raised it again.

And that's when my strategy for dealing with them changed. I ignored them, and focused on everyone else. Suddenly they were irrelevant. When there was something the group said that was hugely inaccurate, I would decide if it was worth jumping in to correct it, but mostly it just wasn't.

It's far too easy to make it worse by accident, so sometimes the best advice is to keep your fingers away from the keyboard. It's much more effective to just tell everyone else what is correct and ignore the trolls.

I wasn't interested in talking to a very small group when I could speak to the country, they were the people that matter and by rising above it, more was achieved. They were only ever worth one reply.

That doesn't mean you can't engage with this highly emotional group, you can, but be smart about it. And when you are you can possibly turn them.

So in this little campaign group, there was an argument with one of them wanting to promote one message, and the others not. This gave me a way in. I followed what they were saying to each other, and stepped in at the right moment, but only talking to the one being singled out.

Suddenly, this one person was receptive to a more calm and logical approach to escape the anger from their own group. I became a voice of reason, pouring oil on their troubled waters. They listened and took on the truth. They even asked if they had been lied to by the campaign group, and being mindful, I said I don't know what they've been told, only what I know is true.

I then watched this one person turn on the group. They reached out to those they were talking to and told them the truth. This person became our advocate, just by being the voice of reason during a storm.

That's how you break through to the anti-vaxxers and intelligent designers. It may not be as quick as you want it to be, but no one leaves a religion overnight. The road to Damascus is a long one with new revelations at every step.

Pick your moments because now your emotions become the calm soothing voice that eases their anger towards you.

This won't always work, and to be honest you won't be able to use it very often, but remember that divide and conquer is as old as the Romans. And what have the Romans ever done for us?

The other type of troll, the one just trying to get a rise out of you through insults and so on, they don't really matter one way or the other, so just ignore them. All you need to do is have a look at their account and see what they are posting.

If all they have are posts trolling people, insulting them and just generally being a douche, then mute them (a better option than blocking), or if they are abusive, block and report them.

You must report them for all abusive behaviour towards you. They'll only set up another account if removed, same if you block them, that's why I like muting.

It's like sticking them in a soundproofed room by themselves to make as much noise as they want.

"What do we live for, if not to make life less difficult for each other?"
George Elliot

4.11 Trolling others via science pages on social media

I've seen a lot of science social media accounts (even follow a couple), where the main purpose seems to be 'look how clever we are and stupid you are'. Quite often they are funny, sometimes quite clever.

But do you remember me saying a while ago about the postgrad students laughing at the questions about intelligent design? Well that's what those pages are.

Science communication is about speaking to the public on their level, not insulting them. The second you take this approach, you do not only yourself, but also what you represent a disservice.

Sure you can follow the accounts, but don't share them, as your non scientist friends will think you're saying they are stupid. This is the equivalent of a comedian loosing them room. You are losing the argument before it has even begun.

As a scientist your deeply held *logical* beliefs can change in seconds when you are presented with evidence proving you are wrong. This is cold hard logic doing the job it was created for.

Deeply held *emotional* beliefs need to be changed over a much longer period of time, in a slow progression.

You chip away at beliefs slowly through the gradual introduction of new information. That's what worked with the person who left the campaign group. The other approach didn't help our case against the coastline person, in fact it worked the opposite way.

And that's what these funny memes are doing, making them angry and digging in deeper into their beliefs. You can say the sky is blue, but they won't want to believe you anymore.

"It's the job that's never started
as takes longest to finish."
J. R. R. Tolkien

5 - Final Thoughts

So that's it, I've given you a whistle stop tour of media theory and given advice on how to communicate, most of which is get out of your head once in while. But not in the good way.

Hopefully I've made a good enough case in how I've presented this advice, and you'll take away some of it to help you in future. I also really hope you read Randy Olson's book next if you haven't already, and if you have, read it again. I can't stress how important the emotional side is in communicating science.

The best ones do it instinctively, Neil deGrasse Tyson, Brian Cox, Jim Al-Khalili. If you've forgotten this, then watch Poppy from Poppy Presents, she has bags of it. And in case you're thinking that her audience is for children, I saw a fantastic zoology presentation to adults by someone who normally only gave this presentation to children.

I watched 50 adults hang onto his every word, and as the projector wasn't working, they all leant forward to look at his iPad to see the images on it. These were adults who were also drinking beer in the Science Museum at the Lates event, so there was a lot of communication noise and real noise to distract them.

It's emotion that people connect to and what connects people. If you add emotion to your communications, you'll be an unstoppable force.

What I've done here is create some rules, or more accurately some guidance for you to use and help guide you through the process when you're doing

#SciComm. As always, rules like these are made to broken, but only when you understand what they really mean.

"Rules are for the obedience of fools and the guidance of wise men and women." Harry Day

Guide/Rules – or what you need to take away from this

1, Stick to one main fact, with three sub facts only in a sentence, don't make it hard to understand

2, Never use just text when you can include a picture

3, Never use a picture when you can use video/gif

4, Never use a long video (more than 15/20 seconds on Facebook and Twitter), unless it's really good and you know they will watch all of it. YouTube is for longer videos.

5, Don't use blog links if you can post text

6, If you have to post a lot of text, put it in an image, people are more likely to read it.

7, Keep the length of text down, subedit, subedit then subedit again.

8, Have a plan for handling trolls

9, BE ENTERTAINING!

10, Follow the formula: **E=MC SQUARE**

E - Einstein "If you can't explain it simply, you don't know it well enough" (It's not about dumbing down, it's about making it understandable).

M - Media Platform (where are you doing SciComm, as this relates to how you present it. Is it a talk, a blog, a vlog, Facebook, twitter, TV, Radio etc. Each has a different style, you need to understand this, or find someone who can adapt what you've already created.)

C - Culture (think about the audience and the culture they come from. This isn't just about the country they are from, but also their social, economic and educational groups)

Square (This relates to culture, as scientists are often seen as geeks/nerds. Think about how you come across to the audience and remember that people relate to people like them. Don't create a them and us feeling by trying to be too clever)

Further Reading

All of the publications listed here are will help you in some way, and I'm including a short explanation for each one to help you decide which will be useful to you.

Understanding media in a global world

Mad, Bad and Dangerous – the scientist and the cinema (2006) – Christopher Frayling
- This book is a really good examination of how film has helped create a stereotype of scientists for the public, and also how it has changed over the years.

Why Scientists are media dumb – Michael Crichton
- There's no easy way to say this, but you need to read the transcript for this speech from way back in 1999. Search for the above title online or use -
http://www.abc.net.au/science/slab/crichton/story.htm

Understanding Media (1964) - Marshal McLuhan
- To explain how important to media theory he is, all I should need to say is he had his own Google doodle recently.

Which Lie Did I Tell? (2000) – William Goldman
- Although this is a follow up to his earlier book about Hollywood, this is still well worth a read. He talks a lot about creating story structure for the films he's written.

The Myth of Media Globalisation (2007) – Kai Hafez
- A very good read that explains the web isn't always the answer when trying to talk to everyone in the world, and that Globalisation is not the all-

persuasive force many believe it to be.

Communications and culture

How Nike Adapts 'Just Do It' To Work Across Cultures (2016) – WARC
- I've included this one here, but you need to have access to WARC to be able to read it. I've summarised it in here, but if you can access it, it's worth a read.

Communications between Cultures (2016) – Edwin McDaniel, Carolyn Roy, Larry Samovar, Richard Porter.
- This book literally does what it says on the tin, and talks about how to understand the differences when trying to reach across cultures.

Geert Hofstede Cultural Differences
- This website allows you to compare one country to another to see what cultural differences and values they hold. Search for the above title online or use https://geert-hofstede.com/countries.html

Global marketing and advertising: understanding cultural paradoxes (2009) – Marieke de Mooij
- Another very worthwhile book to have a look at. When you're communicating science to the public, you are marketing and advertising what you do.

Testing and peer review and practical advice for communications

Don't be such a scientist – Randy Olson.
- Read this book now. That's all I have to say, it's that important for science communications, that if you're serious about doing it you must read it.

On Acting (1987) – Laurence Olivier
- In here he has a lot of advice on how to be on stage, and of course, when you're presenting a talk this will be useful to you. Plus of course, this relates to role theory and the greatest actor of the 20th Century knows a thing or two about that.

Steve Waddington
- Mr Waddington used to be the President of the Chartered Institute of Public Relations, is also a visiting Professor in practice and Newcastle University, and his website gives lots of practical advice. Search for 'Steve Waddington' or use http://wadds.co.uk

Social Media Today
- This website is updated almost daily and gives lots of advice in all sorts of areas for social media and marketing. Often it is a great source of ideas for new practices and also current thinking. Search online for 'Social Media Today' or use
www.socialmediatoday.com

Social Media Playbook
- This is a very useful tool for people new to social media for marketing and PR. The original guidance I wrote for the Coastguard was used to create the first version and lifted sections directly from it. Search for 'GDS Social media playbook' or use
https://gdsengagement.blog.gov.uk/playbook/

Digital Film Making (2007) – Mike Figgis
- Mike Figgis is an Oscar winning director and this books explains in easy to understand terms how to make and edit a video. Well worth a read to help you with your video productions skills.

Vimeo Video School
- These are free to view videos which explain everything from how to hold the camera, how to frame the shot and even the editing process. Search online for 'Vimeo Film School, or you can use either
https://vimeo.com/videoschoolvideos or
https://vimeo.com/blog/category/video-school

Introducing Science Communication: A Practical Guide (2009) – edited by Mark Brake and Emma Weitkamp
- Having read a lot of science communications books as part of my research, this one is the only I would recommend. Once you've read and understood this one and Randy Olson's, it's a good one to read.

The Why and How of Science Communication – Dr Karen Bultitude
- This is also a worthwhile read. Search for the above title or use
https://www.ucl.ac.uk/sts/staff/bultitude/KB_TB/Karen_Bultitude_-
_Science_Communication_Why_and_How.pdf

Filming and Editing Script for video (example)

This is a very simple shooting and editing script for the video I talked about in section 4.7. I didn't use it at the time as there was no need to for such a simple film, but here I am using it to explain the idea. By writing it down in a structured way, you're able to work out what to film on the day and it makes life easier if there are problems.

Shot list	Description Scene / GFX	Additional Info	Run Time
1	Working at a screen, talking on the radio / phone headset. Receiving a fake distress call.	Try to get the badge on the wall, or big screen if poss. GFX on screen with standard emergency message	.05 secs
2	Talking to the camera saying the message	Frame shot to make it look busy. Warn everyone before filming starts.	.10 secs
3	Working at screen, this time sending a message help is on the way	Same as shot 1 GFX on screen with standard emergency message	.05 secs

The next one is more complex, and before you can write the shooting script, you would have to know exactly what you want to include.

GFX of course is short for graphics, which can be either a banner, and image on display or an animation. Notes in italics are me explaining what I mean in the box.

Shot List	Description Scene/GFX	Additional Info	Run Time
1	Intro GFX		.03 secs
1.1	Walking to camera in a lab. Intro from script	Make sure those in the lab know what is going to happen. (*This is important, always consider the health and safety of those where you are filming*)	.10 secs
1.2	Pre recorded VT with VO (*Using what you've already filmed or will film, the VO means voiceover so you have to record what is being said*)	Adjust audio levels in the edit. Poss record separately.	.20 secs

2	Interview in office/lab depending on location	Remember your framing and get filler shots/nods (*Make sure nothing is growing out of heads and also video of yourself nodding to what they say*)	.30 secs (*Interview length for edit*)
2.2	GFX/Animation explaining what they are saying. VO interview	Shot extra video in case it's not ready in time.	.10 secs
3	Facing camera, winding up		.07 secs
4	Exit GFX		.03 secs

So this one is almost news like, but again I'm keeping it simple to show the concept, rather than the details.

The one I used for the Science Week vlogs, which lasted around 90 seconds had over 30 sections and most of that was speaking to camera. Breaking it down this way helped me, and meant I knew exactly what I would be filming and when, what would be on screen when they were talking and that all my framing was correct.

You won't always need to do that if it's just you making the video as you'll know what you want to do. But if you're working with others, it will help them understand what is happening next.

You'll find many examples of these online, but I

always find that keeping it simple is best. And of course writing down in some means you have a plan of some kind, even if it does change as you're making it.

About the author

The first thing the author would like to know is how much he isn't enjoying writing this section. After spending too long wasted in a job he didn't like, he made the change to work in media and hasn't looked back since. After working in PR/Marketing he decided to go back and finish his degree to focus on science communication and is working towards getting a PhD in it. At the time of publication, he is doing a Masters in PR and Multimedia Communication.

Printed in Great Britain
by Amazon